Toxic Chemical and Explosives Facilities

Toxic Chemical and Explosives Facilities

Safety and Engineering Design

Ralph A. Scott, Jr., EDITOR
Department of Defense Explosives Safety Board

Based on a symposium sponsored by the Division of Chemical Health and Safety at the 176th Meeting of the American Chemical Society, Miami Beach, Florida, September 11–13, 1978.

ACS SYMPOSIUM SERIES 96

AMERICAN CHEMICAL SOCIETY
WASHINGTON, D. C. 1979

Library of Congress CIP Data

Toxic chemical and explosives facilities.
(ACS symposium series; 96 ISSN 0087-6156)

Includes bibliographies and index.

1. Explosives—Safety measures—Congresses. 2. Chemicals—Safety measures—Congresses.
I. Scott, Ralph A., 1930- . II. American Chemical Society. Division of Chemical Health and Safety. III. Series: American Chemical Society. ACS symposium series; 96.

TP295.T69 614.8'32 79-9760
ISBN 0-8412-0481-0 ASCMC 8 96 1–352 1979

PRINTED IN THE UNITED STATES OF AMERICA

ACS Symposium Series

Robert F. Gould, *Editor*

FOREWORD

The ACS SYMPOSIUM SERIES was founded in 1974 to provide a medium for publishing symposia quickly in book form. The format of the Series parallels that of the continuing ADVANCES IN CHEMISTRY SERIES except that in order to save time the papers are not typeset but are reproduced as they are submitted by the authors in camera-ready form. Papers are reviewed under the supervision of the Editors with the assistance of the Series Advisory Board and are selected to maintain the integrity of the symposia; however, verbatim reproductions of previously published papers are not accepted. Both reviews and reports of research are acceptable since symposia may embrace both types of presentation.

CONTENTS

PREFACE

Since World War II the development and use of new chemicals, chemical processes, and chemical products have created a new industrial environment. Technology, which gave us progress, also has supplied some techniques for identifying those substances that are hazardous and for protecting workers who use the hazardous materials. To understand the threat posed to employees by explosives and toxic substances, it is necessary to understand the proliferation of chemicals, the difficulties of identifying illnesses caused by exposure to toxic substances, and the ways of controlling and quantitatively measuring these hazards so that the risk to employees is minimized. The establishment of multilevel regulatory controls depending upon the estimated degree of exposure risk and the amount of toxicological information available is presented.

The symposium papers provide a means for developing and subsequently implementing relevant engineering design criteria which are the most efficient and cost effective for explosives and toxic chemical facilities. For both facilities, the series of papers provide the means for classification, measurement, and control of inherent hazards. Practical examples are provided on specific work practices and engineering controls for propellant and propulsion facilities, and for the production, storage, maintenance, surveillance, and demilitarization of explosives. Practical examples of the engineering design criteria used in the design of new university chemical laboratories, in handling and transportation of toxic chemicals, and in the National Cancer Institute's Chemical Carcinogen Facilities are included in this symposium series. This symposium series also provides design criteria needed for both explosives and toxic chemical facilities, examples of which are the design of lightning prediction and protection systems, and electrical requirements in hazardous locations.

Department of Defense
Explosives Safety Board
Alexandria, Virginia
December, 1978

R. A. Scott, Jr.

1

Safety Design Considerations in Munition Plants Layout

RICHARD M. RINDNER and IRVING FORSTEN

ARRADCOM, Dover, NJ 07801

Criteria and methods based on results of accidental explosions have been used until the mid-sixties for the design of high explosive manufacturing and storage facilities. These criteria, however, did not include a detailed or reliable quantitative basis for assessing a degree of protection afforded by the protective facility, and as a result Picatinny Arsenal (now part of ARRADCOM) in the early 60's entered into a broad tri-service program of analysis, testing, and evaluation of structures designed to afford protection against the effects of accidental explosions. The experimental work involved model and full scale testing of reinforced concrete structures and their components. New designs were conceived and the threshold capacities of various structural configurations were determined. The validity of the use of scaled model testing to replace full scale tests was demonstrated.

The product of this 8 year systematic study was the publication of the safety design manual entitled, "Structures to Resist the Effects of Accidental Explosions" (Army's Publication TM5-1300). An outline of studies leading to publication of this manual is shown in Fig 1. The manual contains procedures, charts, and tables required to establish the environment of an explosion and its output in terms of blast and fragments. The relations are presented in such a manner that the type of protective structure may be selected, analyzed, and designed to provide a safe level of protection for personnel, equipment, and for separation of potentially mass detonating materials.

In the course of the application of the manual to the Army-Wide Munition Plant Modernization Program, potential areas for improving and refining the manual appeared. Whenever specific information was not available, the most conservative approach was used. Consequently an extensive program, including testing, was initiated to establish data and procedures to supplement and/or modify the existing regulations and to assist designers in developing the most economical and safe facilities.

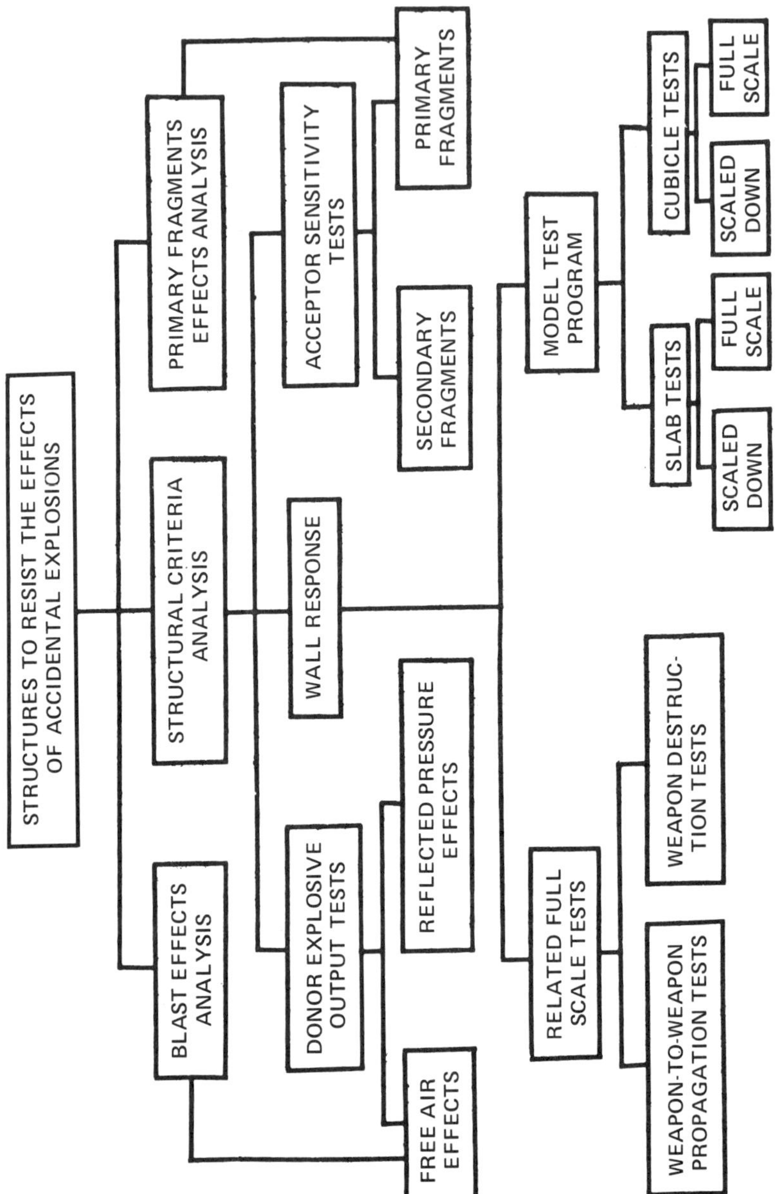

Figure 1. Outline of studies leading to TM5-1300

The overall program was divided into several separate but interrelated phases, which will be presented in the following discussion as shown in Fig 2. These are:

(a) TNT Equivalency Investigation

(b) Safe Separation Distance Determination

(c) Explosive Sensitivity to Impact by Primary and Secondary Fragments

(d) Blast Effects and Structural Response Studies

(e) Hazard Classification Studies of In-Process Hazardous Materials

(f) Development of Special Purpose Water Deluge Systems

In the following pages the individual phases will be discussed in some detail.

TNT Equivalency Study

The purpose of this study is to generate peak pressure and impulse data on explosives, propellants, and other hazardous materials which are compared to similar parameters obtained from a hemispherical surface burst of TNT (Fig 3). The results are reduced to a TNT equivalency value, which is defined as the weight ratio of TNT to test material for a given output condition. Various factors influence the magnitude of TNT equivalency. These include; charge geometry, critical mass/dimensions, confinement, distance from the charge burst, and method of initiation.

Measurements of airblast overpressure and impulse were made at 12 gage locations along a double blast line (Fig 4). The gages were spaced at selected scaled distances ranging from approximately 2 - 20 ft/lbs$^{1/3}$. The pressure transducers were installed flushed with the top surface of a concrete slab in mechanically isolated steel plates. The test item was placed on a steel witness plate located on the surface of the slab. Fastax motion pictures were taken of all tests.

On the basis of experimental data on a variety of explosives, propellants, and pyrotechnics, we have observed that these materials fall into two categories, which can be described in terms of their TNT equivalency-distance curves. The two categories are characterized as marginal explosives and high explosives. The shape of these curves for materials that we call marginal explosives, such as Black Powder, can be seen in Fig 5. The TNT equivalency increases with scaled distance to approximately 10 ft/lb$^{1/3}$ and then decreases. In all cases, however, the maximum value of TNT equivalency is well below 100 percent. Materials,

OBJECTIVES

- SUPPORT MODERNIZATION & EXPANSION PROGRAM
- MODIFY OR SUPPLEMENT SAFETY MANUAL & DESIGN GUIDES

PURPOSE

- UPGRADE PROTECTIVE STRUCTURES, PROCESSES & FACILITIES DESIGNS AGAINST ACCIDENTAL EXPLOSIONS

PROJECTS

- TNT EQUIVALENCY TESTS
- SAFE SEPARATION DISTANCE DETERMINATIONS
- PRIMARY AND SECONDARY FRAGMENTS INVESTIGATIONS
- BLAST EFFECTS AND STRUCTURAL RESPONSE STUDIES
- IN-PROCESS MATERIAL HAZARDS CLASSIFICATION
- DELUGE SYSTEM DEVELOPMENT

Figure 2. Protective technology for accidental explosions

OBJECTIVE

- DETERMINE EXPLOSIVE OUTPUT (PEAK PRESSURE & IMPULSE)
  - EXPLOSIVES
  - PROPELLANTS
  - OTHER HAZARDOUS MATERIALS
- TNT EQUIVALENCY VALUE = $\frac{\text{WT OF TNT}}{\text{WT OF MAT'L *}}$ (%)

  * GIVES SAME EXPLOSIVE OUTPUT AT SAME DISTANCE AS TNT

AIRBLAST CHARACTERISTICS

- DEPENDENT UPON MANY VARIABLES
  - SPECIFIC FORM OF MATERIAL
  - QUANTITY (CHARGE WEIGHT)
  - PHYSICAL STATE (DENSITY, TEMP, CONCENTRATION.)
  - CONFINEMENT
  - GEOMETRY
  - DISTANCE
  - STIMULI (METHOD OF BOOSTERING)

TESTS

- VARIETY OF TEST SETUPS (IN–PROCESS & END ITEM)

Figure 3. TNT equivalency

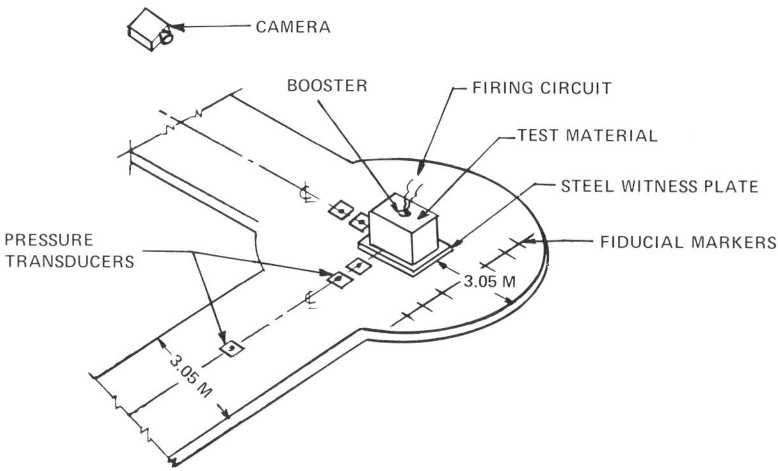

Figure 4. Typical test setup, equivalency tests

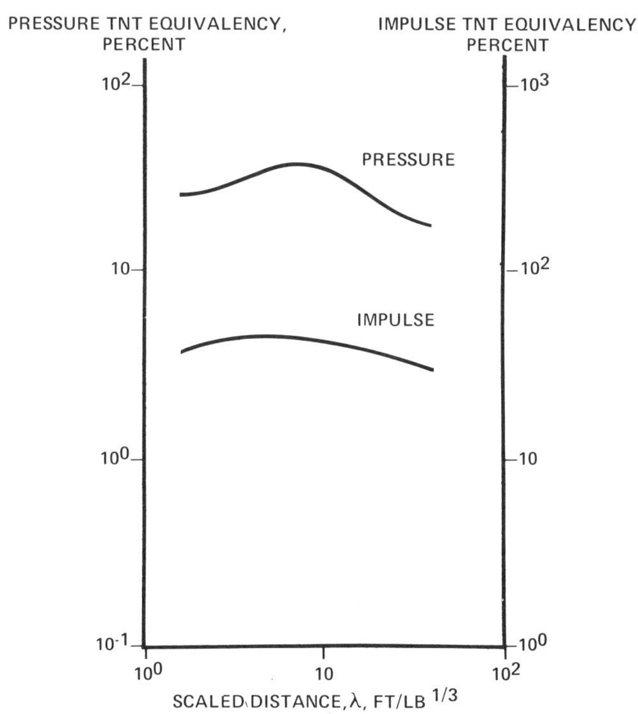

Figure 5. TNT equivalency (black powder weights of 500–4500 pounds)

that we classify as high explosive, such as Nitroglycerine, have a TNT equivalency (as shown in Fig 6) well above 100% at close distances, and decreasing to less than 100% at far-out distances.

Finally Fig 7 summarizes results of the TNT equivalency studies for a few selected explosives, propellants, and pyrotechnics.

Safe Separation Distance Determination

The purpose of this program is to establish safe separation distances relative to explosive end-items and in-process materials, as well as critical and safe depths of bulk explosives on conveyors, hoppers, tubes, or other transfer lines.

Safe separation studies were conducted to achieve increased production and cost effectiveness with improved safety. A typical ammunition production line flow diagram (in this case for the manufacture of 105 mm projectile) consists of several work areas as shown in Fig 8. ((1) Receiving and storage, (2) Box open and inspect (3) Melt Pour (4) Cool (5) Hold (6) Funnel Pull and (7) Riser Preparation.) Explosive material is transferred by automatic conveyor between these work areas. The requirement was to establish safe separation between explosive boxes, pallets with and without funnels, buckets, and to determine critical height of continuous feed flake Comp B and TNT. The objective of these tests was to establish minimum nonpropagation distances between these items so that an explosion chain reaction will be prevented.

The next few photos illustrate the test set-up for various test configurations simulating, as closely as possible, the plant operational line. Fig 9 shows a test set-up for establishing critical height of Comp B flake utilizing a commercially available corrugated rubber conveyor. Fig 10 shows a typical set up for establishment of safe separation distance between pallets containing sixteen (16) 105 mm shell, and Fig 11 shows a test set-up to establish safe separation between the tote bins transporting 165 lb of Comp A7 in a tunnel structure simulating a plant tunnel or ramp. The tests consisted of an exploratory phase to establish the safe non-propagation distance, and a confirmatory phase to confirm statistically the validity of the exploratory test results. Along with the safe separation distance criteria, an attempt is made to evaluate, by statistical analysis, the probability of an explosion propagation occurring. The probability of the occurrence of an explosion propagation is dependent upon the confidence level involved and has a lower and upper limit. The lower limit for all confidence levels is zero; whereas, the upper or practical limit is a function of the number of observations or acceptors tested. Fig 12 represents a family of curves relating the number of tests to the probability of the occurrence of explosion propagation for acceptable levels of confidence.

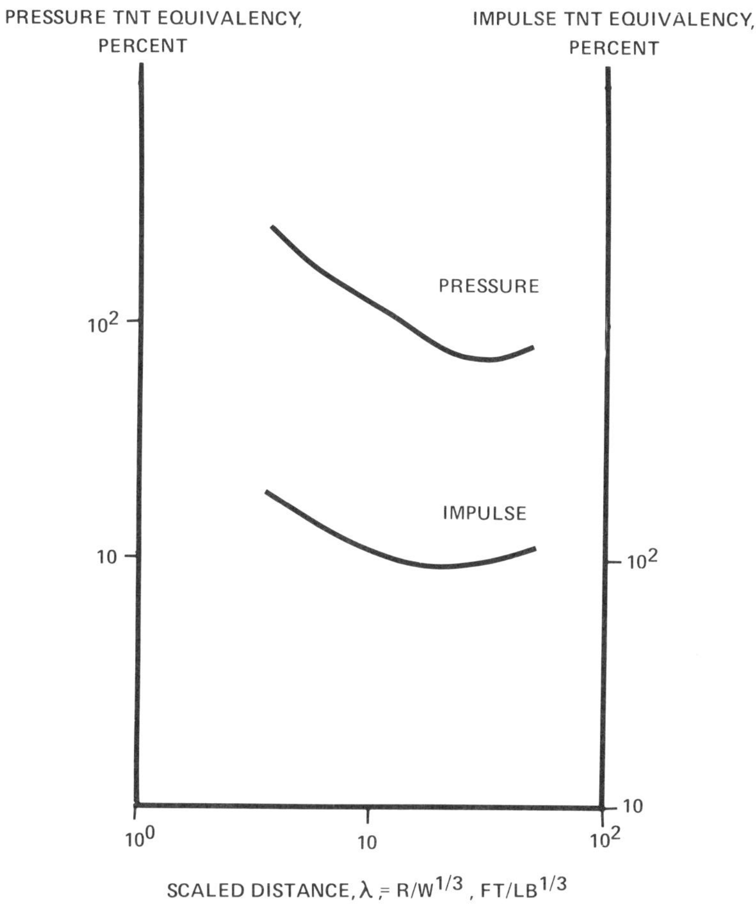

Figure 6. TNT equivalence of nitroglycerine

MATERIAL	IMPULSE EQUIV MAXIMUM (%)	PRESS EQUIV MAXIMUM (%)
BLACK POWDER		
● LARGE SCALE TESTS	50	40
● SMALL SCALE TESTS	24 (30 RECOMMENDED FOR DESIGN)	11
N–5 PROPELLANT		
● SLURRY (88% WATER)	0	0
● PASTE (30% WATER)	2	4
● PASTE (10% WATER)	70	90
NITROGUANIDINE	80	100
GUANIDINE NITRATE	85	55
GUANIDINE NITRATE REACTOR	55	85
NITROGLYCERINE	190	240
PYROTECHNICS		
● 105MM ILLUMINANT	70	60
● M49A1 TRIP FLARE	80	78
● 105MM FIRST FIRE	100	60

Figure 7. TNT equivalency results

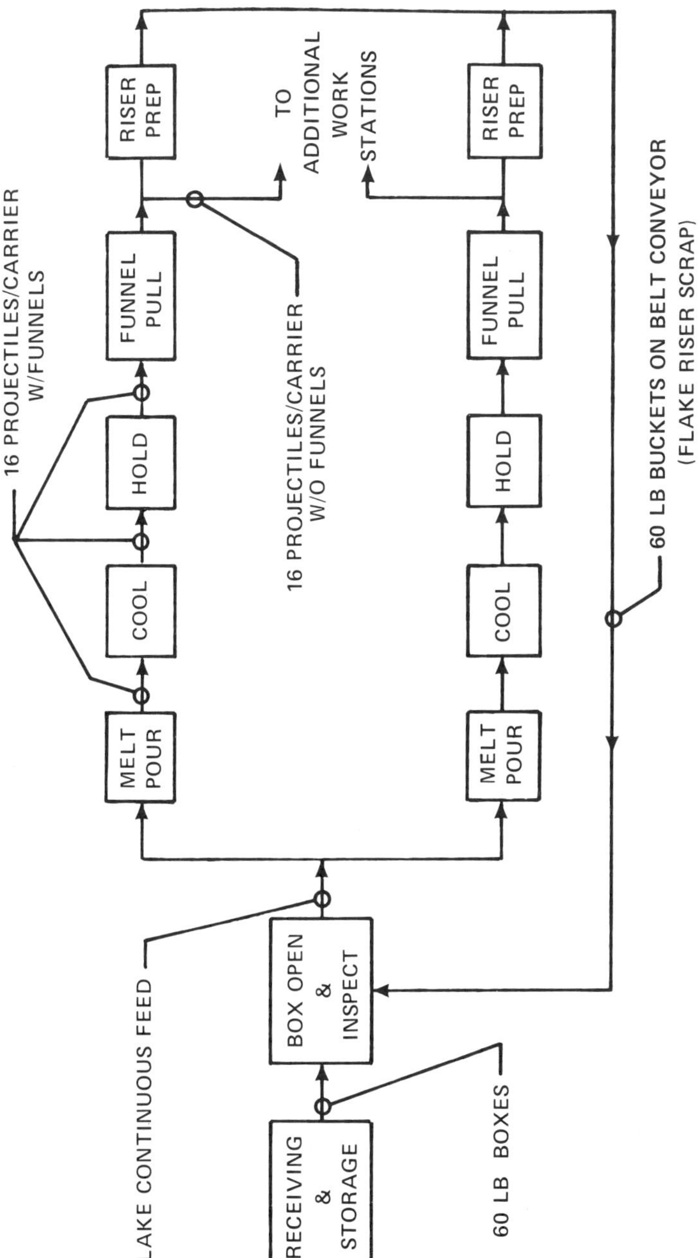

Figure 8. 105mm, M1 projectile melt pour flow diagram

Figure 9. Test set-up to establish crit. ht. of comp B

Figure 10. Test arrangement with 16 projectiles primed

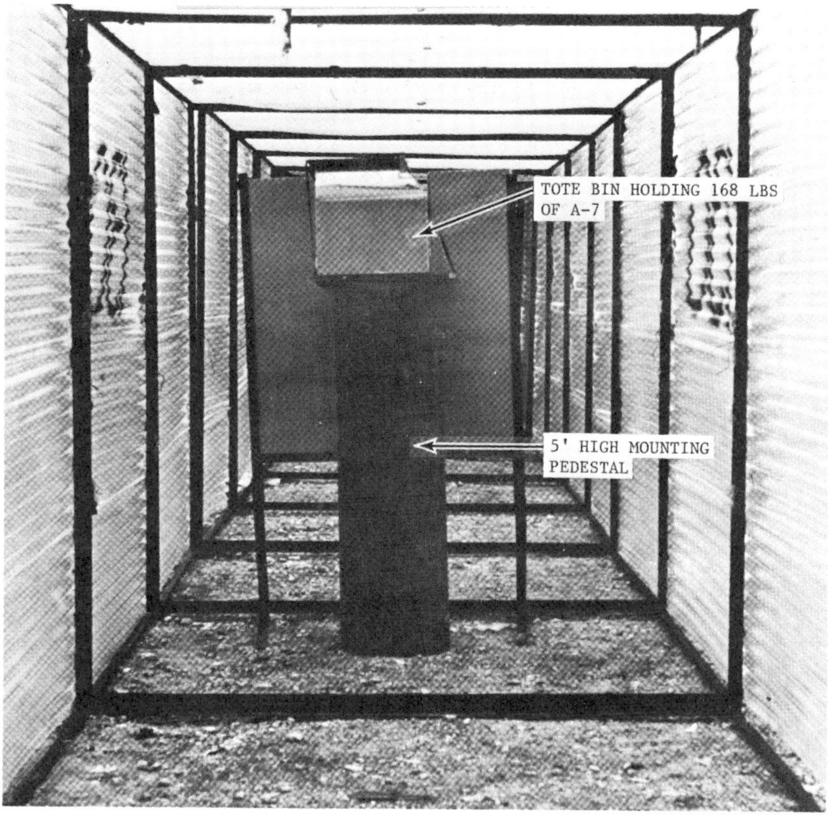

Figure 11. Rear view of acceptor tote bin and celotex witness panels

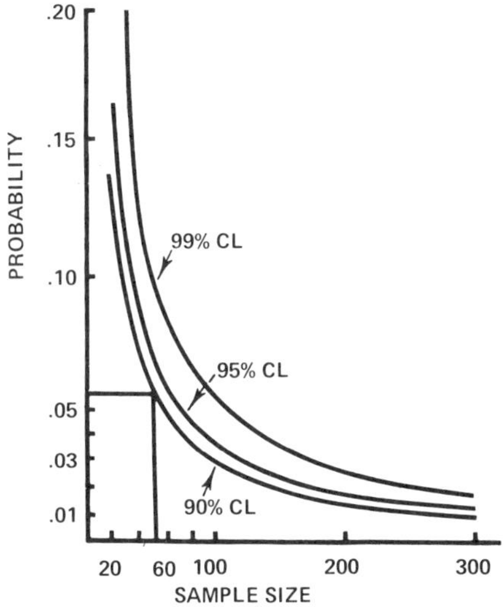

Figure 12. Relationship between probability of propagation and number of tests for a given confidence level

Primary and Secondary Fragments Impact Investigation

The purpose of this experimental program is to establish a fragment mass-velocity relationship below which no detonation propagation will occur.

By definition, primary fragments are those that result from a break-up of the explosive casing in the event of a detonation. Usually these fragments are characterized by having high velocity and being comparatively small in size.

The experimental set-up is shown schematically in Fig 13. The steel fragment impacting the acceptor charge simulates a fragment resulting from the break-up of the shell casing. The entire explosive train, including a booster charge, was placed on top of a 5" square Lucite buffer plate of varying thickness which controlled the fragment velocity. Glued to the Lucite was the steel fragment of desired thickness and frontal area. Two types of targets were used, namely; solid and molten explosives with the acceptor cover plate of varying thickness over the explosive. A high speed camera was the only instrumentation used to record fragment velocity data. Fragment velocity was computed by dividing the distance traversed by the time it took to travel that distance. Fig 14 is a graphical representation for both solid and molten Comp B. As expected, the molten Comp B is more sensitive to fragment impact than solid Comp B, however the difference in sensitivity is not very significant.

By definition, secondary fragments are those other than primary fragments that result from the detonation of explosive charges, such as wall break-up, pieces of equipment, etc. They are usually characterized by having a lower velocity than primary fragments (seldom exceeding 1,000 ft/sec) and having large mass. A series of experiments were conducted at the Illinois Institute of Technology Research Institute (IITRI) Test Facility to determine explosive sensitivity to impact by concrete fragments. Fig 15 simulates a situation where wall fragments resulting from donor detonation impact the acceptor.

The concrete fragments utilized in this program were launched from a 12" gun as shown in Fig 16. This air gun is capable of launching fragments at a wide range of weights and velocities. The experiments utilized both solid concrete cylinders as well as concrete rubble packed into a cardboard container to simulate a fragment from a concrete wall. Typical test results (of the "just filled" configuration representing a 155 mm shell filled with molten Comp B) are shown in Fig 17. As can be seen, high order detonations were recorded at fragment velocities of approx. 1,100 ft/sec and fragment size of 50 lbs. Other tests in the same series investigated sensitivity of variety of items, explosives, and propellants to fragment impact. It is expected that these and future tests will provide empirical relationships between fragment-mass/velocity, for various casing thicknesses and other chemical and physical characteristics of explosive

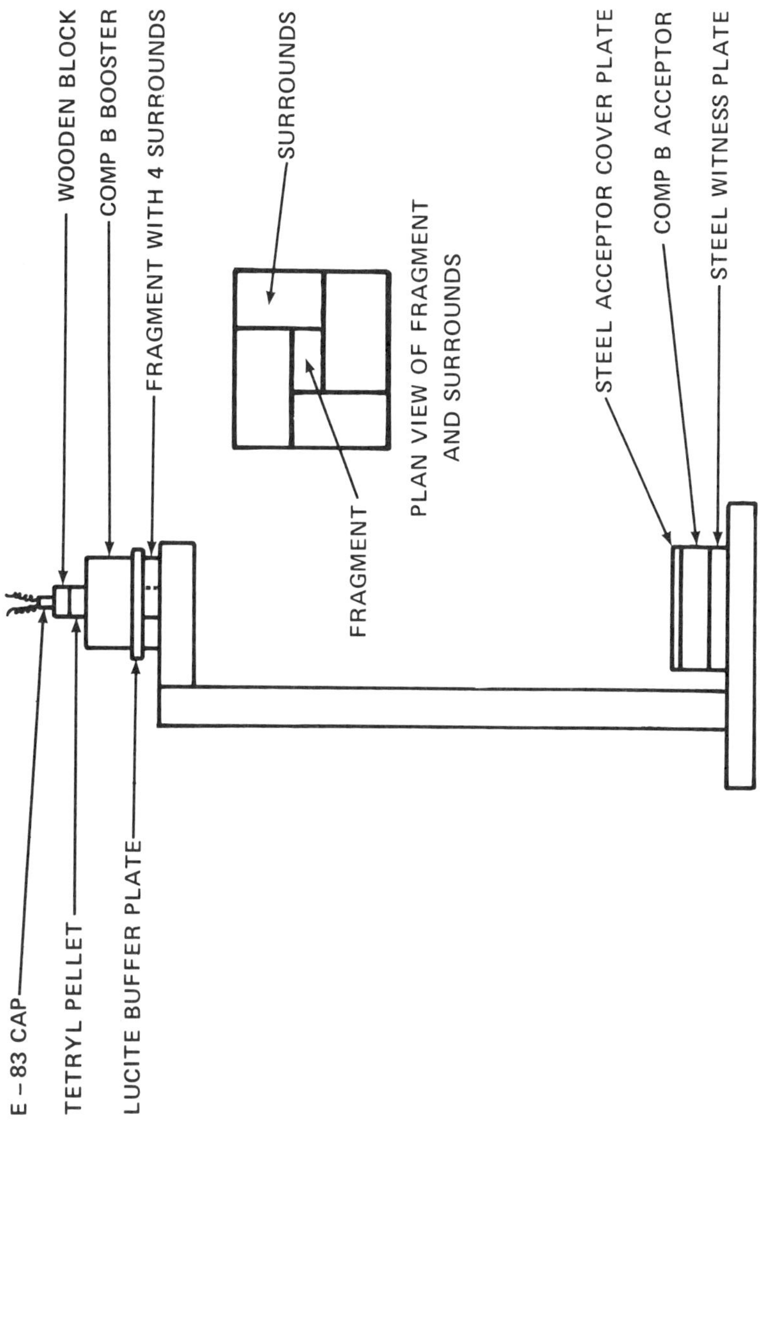

Figure 13. Schematic of experimental set-up

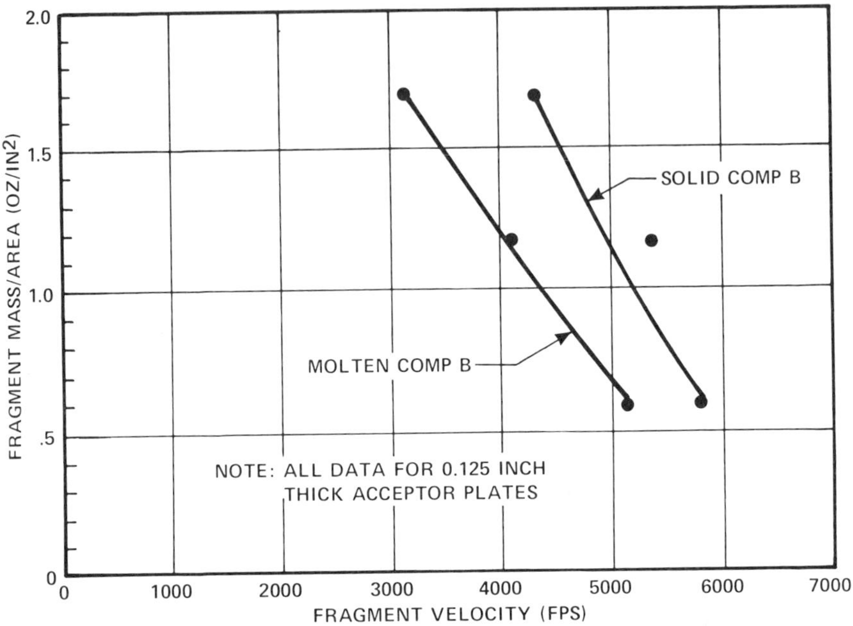

Figure 14. Minimum velocity for detonation—molten and solid comp B

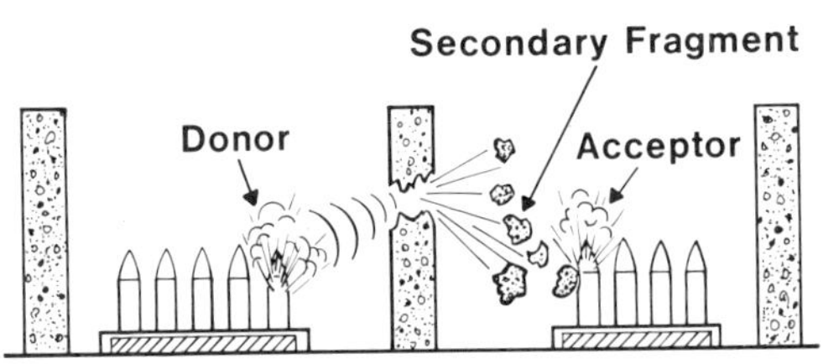

Figure 15. Secondary fragment effect

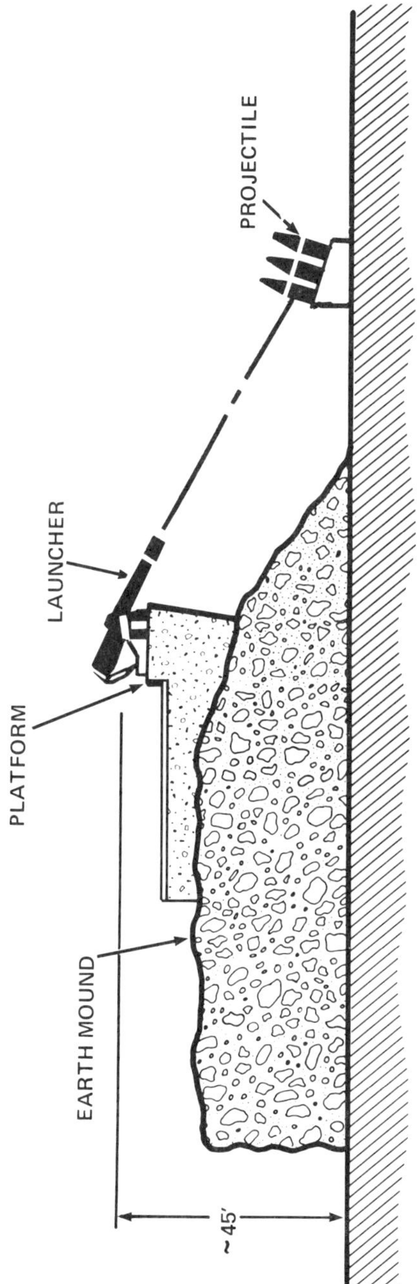

Figure 16. Secondary fragments impact test

JUST FILLED CONFIGURATION

155 MM HOWITZER PROJECTILE, COMP B AT 200° F;
LOADING FUNNEL IN PLACE;
IMPACTED BY 12 IN DIAMETER CONCRETE FRAGMENT

TEST NO	CONCRETE FRAGMENT DESCRIPTION					
	TYPE	LENGTH (IN)	WEIGHT (LB)	VELOCITY (FPS)	KINETIC[a] ENERGY	RESULTS[b]
JS - 1	SOLID	6	70	181	1.2	NO GO
JS - 2	SOLID	6	50	1100[c]	1.5[c]	HO
JS - 3	SOLID	6	50	1100[c]	1.5[c]	HO
JS - 4	SOLID	6	53.5	800	0.86	NO GO
JS - 5	SOLID	6	52	900	1.1	NO GO
JS - 6	SOLID	6	53.5	869	1.0	NO GO
JS - 7	SOLID	6	53.5	1065	1.5	NO GO

[a] ONE ENERGY UNIT = 6.2 X 10^5 FT LB

[b] NO GO = NO REACTION
HO = HIGH ORDER DETONATION

[c] ESTIMATED

Figure 18. Summary of pre-engineered building test results

TEST NO.	PRESSURE (PSI)	DAMAGE
1	0.27	MINOR DAMAGE TO WALL & ROOF PANELS
2	0.55	SIMILAR TO TEST 1
3	0.74	FURTHER DAMAGE TO WALL PANELS & DAMAGE TO GIRTS
4	1.00	FURTHER DAMAGE TO WALL PANELS & GIRTS
5	1.20	FURTHER DAMAGE TO WALL PANELS & GIRTS, & MINOR DAMAGE TO FRAMES
6	1.30	SIMILAR TO TEST 5

Figure 17. Summary of test results

materials. This could lead to a better understanding of the phenomena governing sensitivity and ignition mechanisms for detonation propagation.

Blast Effects and Structural Response for Acceptor Structures

In my introductory portion of this paper I mentioned that in the 60's we developed a safety design manual (TM5-1300) which deals primarily with the design of protective structures located in the high pressure region close-in to a detonation.

At present, the work in this area is directed towards development of design criteria and procedures for acceptor structures located in low and intermediate pressure ranges.

In general, acceptor structures relate to buildings located in pressure range of 10 psi or less. These buildings often contain personnel and equipment which require protection against the blast and fragment output from a donor building where hazardous operations are involved. The selection of the appropriate structural system and materials for acceptor structure design depends on over-pressure level, degree of fragment hazard, contents of the acceptor building and normal operations involving personnel.

It is common in the explosive industry to be separated by either "barricaded or unbarricaded intraline distance" (corresponding to blast loadings of 10 and 3.5 psi respectively) or "inhabited building distance" (corresponding to blast loading of 1.2 psi). These distances are published in the DARCOM Safety Manual (DRCR 385-100) based upon proven scaling laws.

Conventional pre-engineered structures, if used for blast resistant design, would not cover pressure levels required since their capacity to resist blast overpressure (designed for snow and winds loads) seldom exceed 0.2 psi. To design appropriate acceptor structures, tests have been conducted to evaluate the blast capacity of their various components, namely, the steel frames, girts, purlins, side panels and windows.

The blast resistant capacities of pre-engineered buildings can be increased by decreasing the spacing of frames or increasing the size of individual members while still retaining standard pre-engineered building elements. Although these modifications will usually require a cost increase, the added costs are usually more than offset by the cost savings achieved by building separation reduction. A series of tests, adequately instrumented, were performed to verify those modifications, which will produce increased capacity and to identify unknown shortcomings of pre-engineered buildings.

Fig 18 summarizes the pre-engineered building test results including the free-field pressures, frame, girt, and panel displacement, and a brief description of typical damage for each test.

The cold formed steel panels (Fig 19) are widely used for

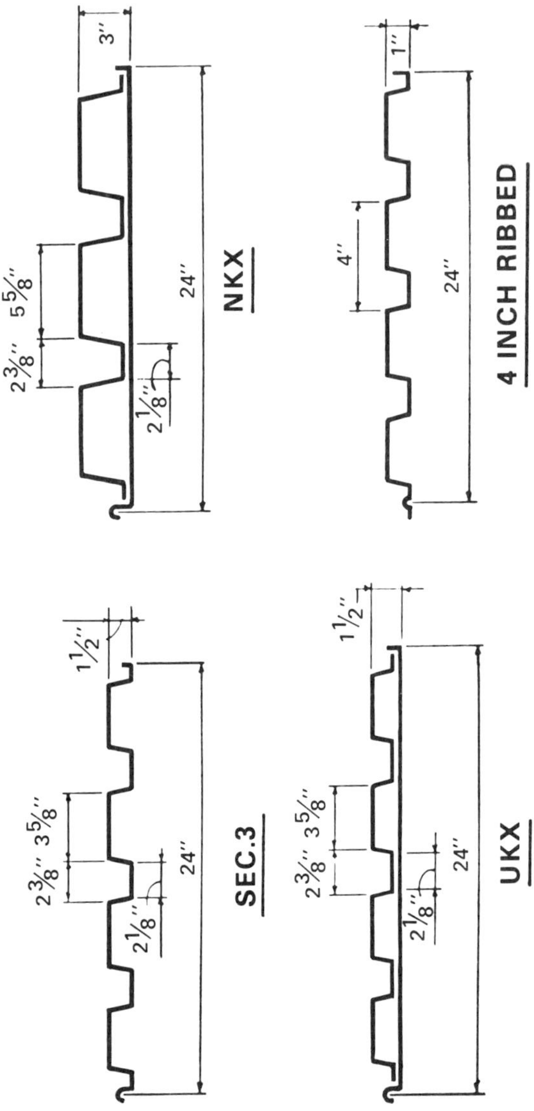

Figure 19. Cross sections of cold-formed steel test panels

roofing, decking, and siding in the construction of steel structures and pre-engineered buildings at explosive manufacturing facilities. Tests were performed at Dugway Proving Grounds, Fig 20, subjecting the panels to overpressures ranging from 0.3 - 1.5 psi which were produced by a detonation of 2,000 lbs of HE. The test results indicated that the panels exhibited considerably greater strength than predicted which utilized the minimum yield strength of about 36,000 psi. However tests indicated that the actual yield strength of the panel averaged about 40% higher than the minimum.

Glass used in the blast resistant structures can be separated into 2 categories. (1) regular glass and (2) tempered glass which consists of regular glass that has been rapidly cooled from its near softening point to increase its mechanical and thermal endurance. Tempered glass is commonly referred to as "safety glass." Tests have been conducted to evaluate the blast resistant capacities of both regular and tempered glass. The results of the tests are summarized in Fig 21. As may be seen from the table the tempered glass can withstand several times the load of regular glass.

Based upon results obtained it was shown, that the use of pre-engineered buildings to provide protection at over-pressure ranges corresponding to inhabited building distances (P_{so} = 1.2 psi) is practical. It should be understood that some building modifications will be needed to insure that the blast resistant capacity of individual building components are consistent. This may require the substitution of larger members used for conventional loads, spacing of members may have to be reduced, and the number of individual components may have to be increased.

Hazard Classification Studies For In-Process Hazardous Materials

Hazards classification is the assignment of a material or an end item (in this case only in-process materials) to a particular hazard class which best describes the threat presented by the material. This requires the use of a hazards classification procedure which provides the guidelines and criteria on which the choice of the hazards class is based. The assigned hazards class of the material is then used as the basis for selecting the proper quantity-distance relationship. Thus, if the hazards classification procedure erroneously assigns a material to the wrong class, either safety is compromised or excessive safety requirements are imposed. Both possibilities are expensive.

The objective of this program is to establish hazard classification procedures, as a supplement to the existing regulatory manual, for in-process materials used during the various stages of propellant and explosive manufacture.

To accomplish this objective, a review of the current hazard classification schemes was conducted. Major deficiencies were uncovered. They related to the classification procedures, to the implementation of the procedures, and to the final usage of the

Figure 20. Damage to sec. 3–20 roof panel

assigned classification (quantity - distance).

As an initial step, the reports of 180 in-process accidents were viewed. A summary of the type of information obtained is shown in Fig 22. The process operation and the probable causative stimuli which led to the accident are given in terms of the number of accidents and the percentage of the total number. Thus the most probable causes of an accident were identified in an accident analysis. The causes varied by process operation and material type. However friction, impact, electrostatic discharge (ESD), and heating were the most commonly identified causative stimuli.

The structure of the hazard classification procedure is shown in Fig 23. The procedure is designed to evaluate sensitivity and effects independently. The sensitivity evaluation will consist of specified tests required for a given process operation. The tests will determine the material's ignition energy which will be compared to the energy developed by the stimuli occurring in the system. The ratio of the material sensitivity energy to the process potential energy will give a safety factor for a specific process operation. A tentative scheme is to classify a material as highly sensitive (Category A) if its safety factor is less than 1.0; as sensitive (Category B) if the safety factor is between 1 and 10; and not sensitive (Category C) if the safety factor is greater than 10.0. Hence, a sensitivity category will be obtained for each stimulus within a given operation. The sensitivity classification will be combined with the effects evaluation to determine the material hazard classification.

The effects evaluation will determine the likelihood of a transition to propagation and the consequences that can occur. Critical height/depth and critical diameter tests will be performed to determine the detonability of a material in bulk or layer form (on a conveyor). Based on the transition results, a decision is made to complete one of the following effects evaluation tests such as: a) firespread test which will include rate of flame spread, heat of flux, and occurrence of fire brands and b) airblast tests including fragment tests.

The results of the effects testing will be used to place the material in a hazard category based on NATO-UN classification scheme and when combined with the sensitivity data will give the material an overall hazard classification. For example, a material which is found to be an intense fire hazard (consequence 1.3) and sensitive (Category B) to initiation by rubbing friction would be placed in class 1.3B.

The results of the tests will be included as a supplement to the NATO-UN hazard classification manual for in-process hazardous materials.

Water Deluge System Application In Munition Plants

A series of projects have been carried out to develop water

GLASS	PEAK PRESSURE (PSI) LOAD DURATION (MSEC)		
	<10	20–40	>100
1/8 IN. TEMPERED	3.0 PSI	2.00 PSI	1.0 PSI
1/4 IN. TEMPERED	6.0	4.00	2.5
3/8 IN. TEMPERED	8.0	6.00	4.0
1/8 IN. REGULAR	0.4	0.25	0.1
1/4 IN. REGULAR	0.7	0.50	0.3
3/8 IN. REGULAR	0.9	0.70	0.5

Figure 21. Blast criteria for glass in rigid frames

PROCESS OPERATION / PROBABLE CAUSATIVE STIMULI	PRESSING	MIXING	REACTOR	CONVEYING	DRYING	FILLING	SCREENING	MACHINING	PERCENT FOR ALL PROCESS OPERATIONS
FRICTION	68	88	17	80	50	74	75	95	59
IMPACT	49	29	8	20	28	35	25	21	32
ADIABATIC COMPRESSION	35			20				5	11
ESD		12			6	22	50	5	8
HEATING	11	6	83	60	61	17		11	26
IMPINGEMENT				20		8	25		3
PERCENT OF ALL OPERATIONS	27	13	9	4	13	17	3	14	

Figure 22. Probable causative stimuli for accidents occurring in specific process operations

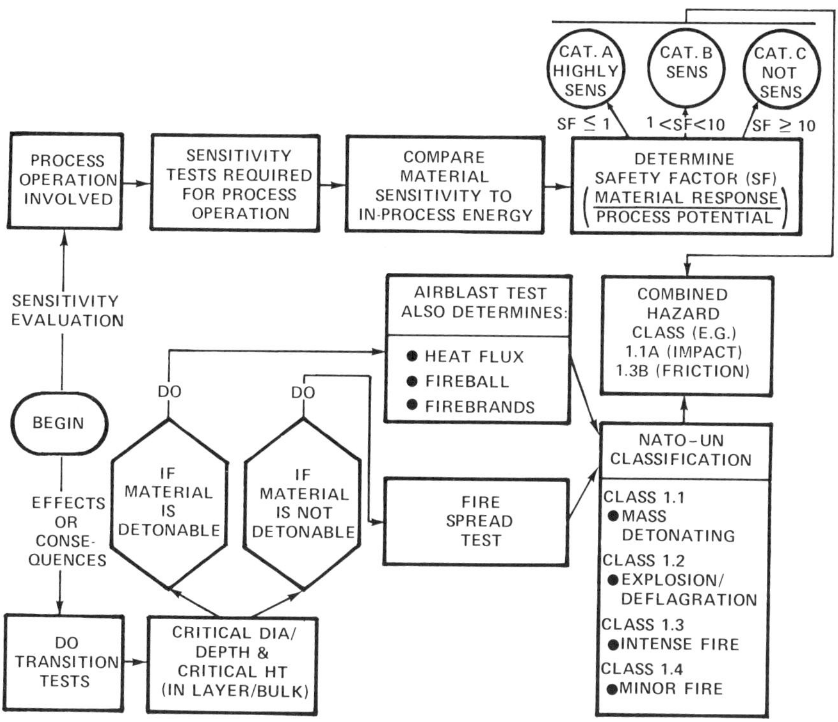

Figure 23. Structure of the hazard classification procedure

deluge fire suppression systems for use by the GOCO plants. Each of the projects has been successful in demonstrating that rapidly activated water deluge systems can be effectively used to suppress and extinguish propellant fires, and in other application can be equally effective to combat extraneous fires as may be caused by a large scale detonation. The purpose of this program was to evaluate the effectiveness of water deluge applied to quenching and extinguishing propellant fires as well as extinguishing fires after explosive detonation.

A major consideration in the design of a deluge system is to determine the type of fire and the rate at which the material is burning in order to establish the probable time in which the fire can be attacked and extinguished before any catastrophic event occurs. A fast burning fire characterized by a large propellant burn, would require a rapidly responding deluge system to control the rate of burn, and to prevent transition of the rapid deflagration into a detonation. Even without a detonation, large propellant fire causes the generation of expanding gases that could cause bulging or explosions of the entire building. Another type fire must also be considered, such as a fire occurring as the result of, or after a detonation. Typically the detonation of a quantity of high explosives will cause severe structural damage to the building and could in some instances have destroyed a conventional overhead deluge system. In area where a detonation can occur, and extraneous fires can start in the same or adjacent buildings, a "hardened" deluge system is required which is capable of surviving the explosive blast and possible fragment impacts, while still being capable of functioning and extinguishing extraneous fires.

Still another consideration is fires occurring in machine equipment such as electrical fires or lubricating oil fires in a fuze assembly machine. Here, primary consideration is the rapid detonation of the fire and deactivation of the equipment to prevent the continous machine operation which would cause the fire to intensify or propagate into a detonation. Fire fighting in this case, in the form of a water deluge system, is the secondary action to extinguish the cause of the fire.

Each of the above problem area studies thus far have required a separate investigation because of the different quantities of material involved, difference in the degree of confinement, differences in the mode of ignition and propellant geometries, and specific constraints dictated by the water pressure and water adequacy problems at each ammunition plant. Thus far, each of the programs has been conducted under great urgency to provide a required design for a unique water deluge system.

An extensive test program using 3 detectors (UV, IR, and Visible) simultaneously with 7 different flame sources, indicated that the UV detectors were more responsive in most cases. Two distinctly different type deluge systems will be discussed as applicable to munition processes. One deals with a "hardened" water deluge system, and the second deals with a "conventional" deluge system. A "hardened" deluge system is one that would be used where a detonation, as opposed to a deflagration, would occur. In order to protect from extraneous fires that could be ignited in the plant as a consequence of a prior detonation, one must use a deluge system that has been hardened against blast and fragments. A conventional deluge system is typified by the overhead water sprinkler system than can sustain intense heat and pressures, but cannot sustain blast and fragment attack. To illustrate the hardened deluge system concept, consider the situation wherein Composition B in cardboard boxes is being transported along a conveyor line. This particular system called for the design of a deluge system that could sustain a blast and subsequent fragments resulting from the detonation of one of the sixty pound boxes of Composition B, yet still be able to function and put out extraneous fires as they might occur. In Fig 24, we see a simulated tunnel and two boxes of Composition B moving along a simulated assembly line representing Lone Star AAP. These two boxes have been set at a previously established 12 ft. minimum safe separation distance. It is assumed that propagation of a detonation will not occur. However, extraneous fires could indeed propagate to the acceptor box of explosive. In this application a UV detector was used to detect the fire. This detector in turn activated a high speed Primac water release valve, thus releasing water from a hardened deluge system. This system was set in the floor beneath the level of the concrete test pad and indeed survived the detonation of the donor box of Composition B. In Fig 25 the aftermath of the detonation of the donor box can be seen. The tunnel and the conveyor system have been totally destroyed, however, the deluge system has not been damaged in any way by the blast and fragments, thus it was capable to act and extinguish any extraneous fires that may have occurred.

Let us now consider the unhardened or more conventional deluge systems. In Fig 26, one element of the 105 mm bag-loading facility at the Indiana AAP is shown. Granular M-1 propellant is loaded into bags, and these bags move down a conveyor line and are stored into large accumulators. In the accumulator room at the Indiana AAP, seven accumulators are parallel to one another, each approximately 70 - ft. long. The total bags contained in that room at any one time can approach as much as 70,000 pounds. Therefore, should a detonation or deflagration occur in one of these accumulators, the results could be catastrophic to the entire line. First and foremost it was demonstrated that fires in these quantities of propellant can be extinguished before the fire transcends into a detonation. The intention of the program in this case was to see if a fire could be put out should one occur in the accumulator. In Fig 27, a donor accumulator and an acceptor (or witness) accumulator are shown. The donor on the right contains bagged propellant 16 in. deep simulating the real plant situation. In the witness accumulator on the left, the bags are

Figure 24. Simulated conveyor line for comp B explosive

Figure 25. Operating nozzle system following detonation of donor box

Figure 26. Indiana AAP accumulator deluge test set-up

Figure 27. Accumulator fire 10 seconds after ignition, water deluge activated

only two layers thick and, should a fire occur in one of the witness acceptors, it would have been ignited from the top as a result of firebrands being erupted from the donor, deflected off the roof and back down onto the acceptor accumulator. To battle large accumulator fires, a deluge system, consisting of two parallel lines, was developed. The first line is a series of high pressure deluge nozzles designed to have force behind the water such as they can penetrate through the bags of propellant and prevent underburn. Parallel to the deluge nozzles a series of area coverage nozzles very similar to a more conventional overhead fire sprinkler system was installed. UV detectors were used to sense the start of a fire and to trigger the water deluge system. In Fig 27, the fire can be seen ten seconds after ignition, flames reaching over 100 ft. in the air, and material raining down on the floor, surrounding the accumulators. These tests clearly demonstrated that a water deluge system had been designed, built, and tested which was capable of suppressing the fire and preventing a detonation and ultimately in extinguishing the fire within eight feet along the accumulator from any random point of ignition.

Program Accomplishments and Future Goals

Finally, I would like to complete my presentation by listing some of our program accomplishments and future plans to further safe-guard hazardous operations both in government and in industry.

1. Program Accomplishments:

   a. Established TNT Equivalency for explosive and propellants

   b. Established statistically valid safe separation distance for munition items and explosive materials simulating actual operating conditions.

   c. Established effects of primary and secondary fragments impact on explosive end-items an in-process materials.

   d. Performed analytical and experimental work to broaden scope of TM5-1300

   e. Prepared a Technical Data Package for water fire protection system for conveyor lines transporting high explosives

   f. Developed tailor-made water deluge systems to extinguish M1 Propellant fires.

g. Provided design assistance to GOCO plants in their preparation of design criteria and concepts for facility modernization.

2. Future Study

a. Investigation of other types of construction and material

b. Establishment of fragment effects from frangible structures

c. Development of quantitative approach to safety problems in explosive storage and manufacturing plants (risk analysis)

d. Development of protective shields for personnel and equipment

e. Development of explosive dust detection system

f. Performance of additional TNT equivalency studies on inprocess and other materials

g. Development of analytical procedures for characterzing primary and secondary fragments

Closing Remarks

It is difficult in the brief time allotted this presentation to describe all our Protective Technology Programs in more detail. However, further details, if desired, can be found in the bibliography inclosed in this paper.

SELECTED LIST OF REPORT ON THE PROTECTIVE TECHNOLOGY PROGRAM

1. R. M. Rindner - Establishment of Safety Design Criteria for Use in Engineering of Explosives Facilities and Operations - Report No. 1 Sympathetic Detonation PA Tech Report DB-TR 1-59 Jan 1959
AD #305775

2. R. M. Rindner - Establishment of Safety Design Criteria for Use in Engineering of Explosive Facilities and Operations Report No. 2 Detonation by Fragment Impact, PA Tech Report DB-TR 6-59 May 59 AD # 303421

3. R. M. Rindner and S. Wachtell - Establishment of Safety Design Criteria for Use in Engineering of Explosive Facilities and Operations Report No. 3 Safe Distances and Shielding for Prevention of Propagation of Detonation by Fragment Impact, PA Tech Report DB-TR-6-60 Dec 1960 AD #321962

4. R. M. Rindner and A. Schwartz - Establishment of Safety Design Criteria for Use in Engineering of Explosive Facilities and Operation Report No. 5 Supporting Studies Through Dec 64, PA Tech Report 3267 June 65 AD #470306

5. R. M. Rindner, S. Wachtell, L. W. Saffian - Establishment of Safety Design Criteria for Use in Engineering of Explosive Facilities and Operation Report No. 8 Supporting Studies, Jan - Dec 1965, PA Tech Report 3484, December 1966. AD #643871.

6. R. M. Rindner, S. Wachtell, L.W. Saffian - Establishment of Safety Design Criteria for Use in Engineering of Explosive Facilities and Operations - Supporting Studies Jan - Dec 1966 Report No. 9 - PA Technical Report 3594 June 1967. AD #818-394

7. E. Cohen, N. Dobbs, R. M. Rindner - Blast Pressures and Impulse Loads Produced by Explosions in Cubicle-Type Structures - PA Technical Report 3604 May 67

8. R. Rindner et al - Establishment of Safety Design Criteria for Use in Engineering of Explosive Facilities and Operations - Supporting Studies - Jan - Dec 1967 Report No. 11, PA Tech Report 3712 Sept 1968

9. S. Levy, R. Rindner, S. Wachtell, N. Dobbs & M. Dede- Summary of Full & Model-Scale Reinforced Concrete Bay Structure Series - PA Tech Report 4168

10. R. M. Rindner and W. Wachtell - Custom Designed Blast Resistant Structures - Presented at the Seminar on "Disaster Hazards" Sponsored by the Central States Section of the Combustion Institute at the Manned Spacecraft Center Houston, Texas, 7-8 April 1970

11. N. Dobbs, M. Dede, R. Rindner - New Concepts in the Design of Structures to Resist the Effects of Explosive-Toxic Detonations PA Tech Report 4060 May 70 (presented at the Seminar on "Disaster Hazards" Sponsored by the Central States Section of the Combustion Institute, 7-8 April 1970)

12. R. Rindner and N. Dobbs - Concept Study of the Proposed Modification of Building 1606 at Rock Mountain Arsenal - Volume I - Results of Concept Study - PA Technical Report 4039 Feb 1970

13. R. Rindner, N. Dobbs - Concept Study of the Proposed Modification of Building 1606 at Rock Mountain Arsenal - Volume II - Concept Study Calculations PA Tech Report 4040 Feb 1970

14. L. Saffian and R. Rindner - Application of New Protective Technology in Design of Explosive Processing and Storage Facilities - Presented at the annual meeting of the "Institute for Chemistry of Propellants and Explosives" - West German Federal Republic - Sept 1970

15. R. Rindner - New Protective Design Technology and Related Explosive Equivalency Studies in the Design of Explosive Processing Facilities - Presented at the 8th meeting of the Integration Committee on Propellants and Explosives - April 1971

16. J. Swatosh, Jr., and H. Napadensky, TNT Equivalency of N5 Slurry and Paste, Illinois Institute of Technology Research Institute (IITRI) TR J6278, September 1972

17. M. Dede, N. Dobbs, R. Rindner, Project Coordinator, Preliminary Estimate of Concrete Thickness and Construction Costs of Laced R. C. Structures, PA TR 4441, October 1972.

18. N. Dobbs, R. Parker, R. Hendershot, Approved Safety Concepts for Use in Modernization of USAMUCOM Installations: PA TR 4429, October 1972

19. J. M. Ferritto, Determination of Blast Leakage Pressures and Fragment Velocity for Fully Vented Protective Cubicles, Naval Civil Engineering Laboratory (NCEL) TR R780, November 1972

20. H. Napadensky and J. Swatosh, Jr., TNT Equivalency of Black Powder, (Vol I & II) IITRI TR J6265-3, September 1972

21. J. Swatosh, Jr., and H. Napadensky, Explosive Hazard Classification of M1 Propelling Charge in its Container, (Vol I & II), IITRI TR J6265-2, September 1972

22. R. Rindner and S. Wachtell, Establishment of Design Criteria for Safe Processing of Hazardous Materials, presented at the 65th Annual AICHE symposium on "Loss Prevention in Chemical Industry" November 1972

23. C. Anderson and R. Rindner, Separation Distance Tests of 155 MM (M107) Projectiles, PA TR 4425, January 1973

24. R. Rindner, J. Swatosh, Explosive Sensitivity of 155 MM Projectile, RDX Slurry, and Black Powder to Impact by Concrete Fragments, PA TR 4594, December 1973

25. H. Napadensky, J. Swatosh, and A. Humphreys: R. Rindner Project Coordinator, TNT Equivalency of Three Pyrotechnic Compositions, PA TR 4628, June 1974

26. D. Allan et al, R. Rindner Project Coordinator, Development of An Analytical Model to Predict Explosive Propagation Between Adjoining Explosive Items, PA TR 4722, September 1974.

27. D. Spandoni, R. Rindner Project Coordinator, Determination of Critical and Safe Heights of M1 Propellant for 105 MM Howitzer Ammunition, PA TR 4754, December 1974

28. G. Tseng et al P. Price, Project Coordinator, Design Charts for Cold Formed Steel Panels and Wide-Flange Beams Subject to Blast Loads, PA TR 4838, August 1975

29. R. Kukuvka and H. Sarette; R. Rindner Project Coordinator, Determination of Minimum Non-Propagation Distance 105-mm M1 Projectiles Grouped Sixteen on a Pallet, PA TR 4869, September 1975

30. J. Healey et al, P. Price Project Coordinator, Design of Steel Structures to Resist the Effects of HE Explosions, PA TR 4837, August 1975

31. W. Stea and P. Price, Overturning and Sliding Analysis of Reinforced Concrete Protective Structures, PA-TR-4921, February 1976

32. G. Petino, D. DeMella, and R. Rindner, Sensitivity of Cased Charges of Molten and Solid Composition B to Impact by Primary Steel Fragments, PA-TR-4975, June 1976

33. D. Kossover and R. Kukuvka, Determination of Minimum Non-propagation Distance of 155mm M107 Projectiles Grouped 24 on a Pallet, PA-TR-5008, September 1976

34. R. J. Odello and P. Price, Grouped Shock Effects from Accidental Explosions, PA-TR-4995, November 1976

35. K. Ghandi and R. Kukuvka, Critical Depth Tests of Comp B Flake, PA-TR-5014, November 1976

36. N. Dobbs, A. Ammar, S. Weissman, and P. Price, Blast Capacity Evaluation of Single Revetted Barricades, PA-TR-5009, November 1976

37. C. H. Johnson and R. M. Rindner, Explosive Tests for Establishing the Hazard Classification for MISP/105mm Propellant in Automated Single-Base Finishing Operations, PA-TR-5020, January 1977

38. R. Arya and P.Price, Blast Capacity of Belowground Structures, ARLCD-TR-77006, May 1977.

39. W. Stea, G. Tseng, D. Kossover, P. Price, and J. Caltagirone, Nonlinear Analysis of Frame Structures Subjected to Blast Overperssures, ARLCD-TR-77008, May 1977

40. W. Stirrat, Critical Depth Tests of Bulk TNT Flake Explosive, ARLCD-TR-78003, January 1978

41. W. Stirrat, Determination of Minimum Non-Propagation Distance of M42 and M46 Grenades Without Fuze, ARLCD-TR-78004, February 1978.

42. H. McLain and W. Seals, Water Deluge Fire Protection System For Conveyor Lines Transporting High Explosives, ARLCD-TR-77001, December 1977

43. H. Napadensky, R. Joyce, R. Rindner, and D. Satriana,. Engineering And Experimental Studies for The Development of Hazards Classification Data on Propellants and Explosives, (to be published)

RECEIVED November 22, 1978.

2

Shielding of Facilities for Work with Explosive Materials

DAVID J. KATSANIS

Chemical Systems Laboratory, Aberdeen Proving Ground, MD 21010

A relatively new concept called suppressive shields is offered to provide protection to the area surrounding hazardous work with pyrotechnic and explosive material. At present, these operations are either limited to small quantities, widely dispersed, or segregated by barricades. Suppressive shields provide an alternative in the form of a vented steel enclosure.

Figure 1 illustrates the concept of a suppressive shield. The enclosure usually consists of a structural steel framework with built-up panels of steel angles, I-beams, perforated plate, or louvered panel. The space between panel components allows gaseous products of combustion to pass through while suppressing flame and, in case of a detonation, reducing blast overpressure to a safe level. There is no direct path through the panel for fragments to escape. All fragmentation effects are confined within the enclosure.

The shields can be any size. They can be small transportable laboratory shields or large structures in a building similar to the concrete barricades sometimes used (1).

Typical reinforced concrete barricades are shown in Figure 2. The barricades are designed as cubicles with three walls to withstand direct blast pressure and to prevent propagation of a detonation from one area to the next (2). They do not prevent hazardous run-up reactions where a fire can start in one cubicle and through pyrotechnic dust in the air or accumulation on equipment, spread from one cubicle to another until the entire facility is in flames.

In the event a detonation occurs, the cubicles do not prevent wide dispersal of damaging primary and secondary fragments, nor do they prevent blast overpressure leakage beyond the open edges. The blast overpressure from the open edges of the cubicle can spread over the outside of the building wall and sometimes is large enough to collapse the frame front wall and roof or the masonary end and back wall. A special reinforced building design can prevent this, but adds considerably to building costs.

Since suppressive shields are full enclosures, they perform

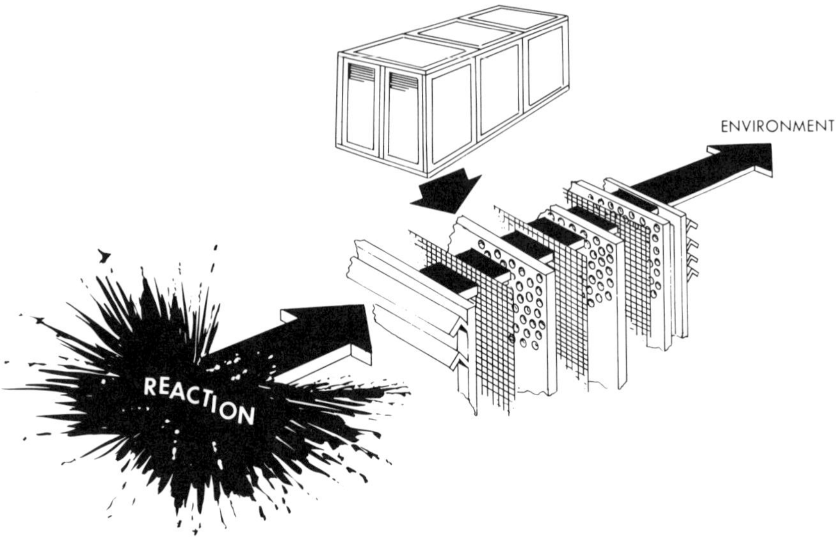

Figure 1. Suppressive shield

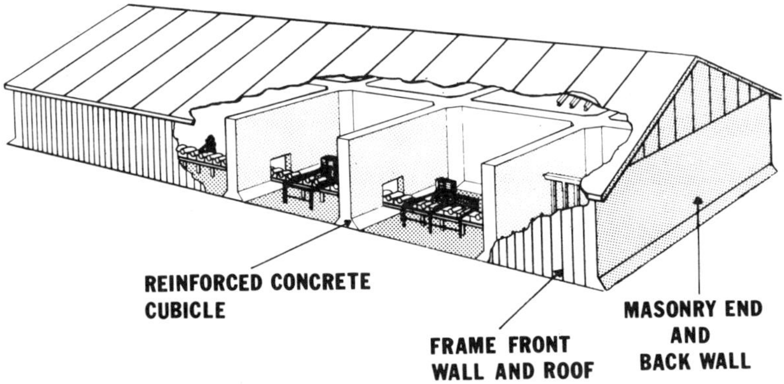

Figure 2. Typical concrete cubicles

in a different way from these cubicles.

Suppressive shields will:

- Confine all fragments from a detonation.
- Attenuate blast pressure to a safe level in all directions.
- Reduce fireball diameter sufficiently to prevent spreading of the fire beyond the local area.

Another particularly attractive feature of suppressive shields is that they are modular in design for quick erection and modification to provide maximum protection and flexibility.

Several general classes of shield designs have been conceived. As shown in Figure 3, some have cylindrical or spherical configuration while others are rectangular frame and panel designs.

Generally the configuration is governed by the dominant hazard, i.e., blast, fragment, or flame. If blast pressures are the factors which are most important in design of the structure, the shield will usually have a cylindrical or spherical shape. The rectangular frame and panel structures are typically used where the dominant hazard is flame.

In this paper, the Group 6 and the Group 5 Shield will be discussed to illustrate details of two typical applications. The Group 6 Shield illustrates a unique spherical design which is small and portable for use with laboratory quantities of primary explosives. The Group 5 Shield demonstrates the modular design concept that makes suppressive shields an attractive alternative to inflexible concrete barricades.

Discussion of Group 6 and Group 5 Shield Designs

Group 6 Shield. The Group 6 Shield is spherical. The requirement for this shield is that an operator be capable of transporting on a push cart small quantities of extremely hazardous primary explosive material. It is not feasible to vent this shield because of the hazardous material involved and the close proximity of the operator.

The two foot diameter spherical steel shell shown in Figure 4 is 1/4 inch thick and weighs about 165 pounds. A rectangular tray is used inside the shield to carry 10 cups of primary explosive material. Each cup contains 70 grams of lead azide in a typical application. Total weight of explosive is limited to 700 grams lead azide or equivalent (1, 3).

One interesting use of this shield will be illustrated. Although it is not a laboratory application, it is interesting because it shows how quantities of initiating or primary explosives can be handled in detonator production without exposing operating personnel to explosive hazards.

Use of the shield and cart shown in Figure 5 will result in about 3 million dollars savings in modernization of the detonator production line at Iowa Army Ammunition Plant. Costly automated

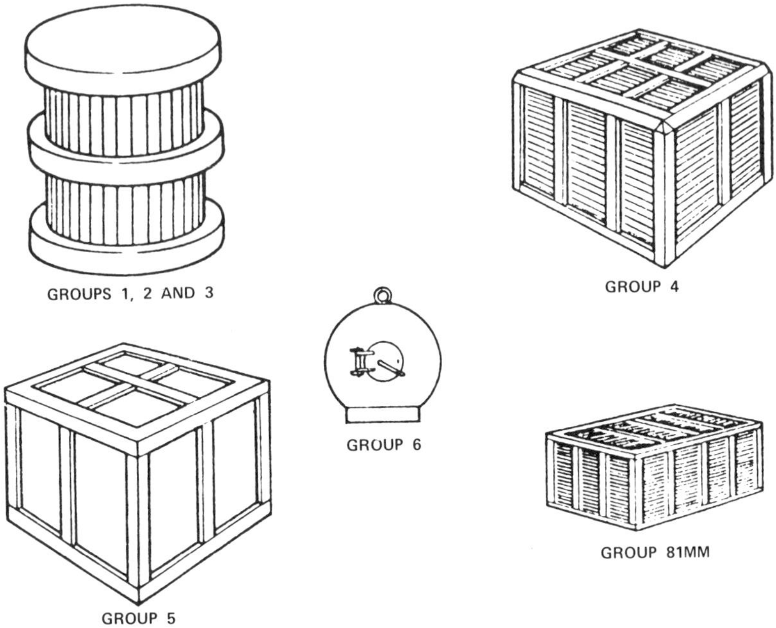

Figure 3. General configuration of suppressive shield groups

Figure 4. Group 6 suppressive shield

Figure 5. Group 6 suppressive shield cart

conveyors are replaced by a manual system which makes use of the Group 6 Suppressive Shield mounted on a cart as shown. The ring at the front mates with an opening in a storage barricade.

A fixture for mechanical function testing is photographed in Figure 6. It illustrates the attachment of the shield to the barricade. The operator rolls the cart up to the barricade where a clamp is closed and the shield is locked to the barricade. Transport mechanisms shown behind the barricade remove the tray with explosive material from the shield. At no time is the operator exposed to explosive hazards.

A series of tests were conducted to verify the adequacy of the design.

Figure 7 is a photograph of the set up for proof test of the system. The Group 6 test shield is supported by a wooden frame and is attached to the storage barricade in the middle of the photograph.

Summary data from proof tests, tabulated in Table I, show that the Group 6 Shield and the storage barricades are adequately designed. The system contained all effects from blast of service charge with a sound level of 146 db outside the shield. One hundred forty-six decibles is roughly the noise level from firing a service rifle such as the M-16 rifle. Ear protection is desirable but a single exposure will often produce no ear damage. This indicates a safe environment for the operator. A charge weight of 307 grams C-4 (37% above the design service charge) was required to rupture the shield. The proof charge which is 25% above service charge caused a bulge in the bottom of the shield but no rupture.

Group 5 Shield. The Group 5 Suppressive Shield is designed for use with pyrotechnic material (4).

Although the example discussed here is a large fixture, these shields can be scaled down in size for laboratory use or modified to meet special requirements.

One of the features that adds to suppressive shield utility is the modular design illustrated in Figure 8. The panels for the Group 5 Shield are laid out by the foundation ready for assembly.

Each panel is about 9 feet long and 5 feet wide. When erected they form a cube about 10 feet long on each side and about 9 feet high. As illustrated by the Group 5 panel section shown in Figure 9, each panel is a composite structure with a double row of interlocking structural steel angle beams arranged as shown in the photograph. There are three perforated plates, one on the inside and two on the outside. Wire screening was added between the panel layers for additional flame suppression, but it proved ineffective. Tests demonstrated that there was sufficient exposed metal surface to suppress flames effectively without the addition of metal screens.

The photograph in Figure 10 shows the panels being moved

Figure 6. Mechanical function test fixture

Figure 7. Proof test fixture

TABLE I. SUMMARY OF PROOF TESTS

MANUAL EXPLOSIVE TRANSPORT SYSTEM

ITEM	CHARGE			FUNCTION
	WEIGHT* (gm)	COMPOSITION	LOCATION	
GROUP 6 SHIELD	708	NOL	CENTER	SATISFACTORY
GROUP 6 SHIELD	567	LEAD AZIDE	10 CUPS	SATISFACTORY
GROUP 6 SHIELD AND STORAGE BARRICADE	567	NOL	CENTER	SATISFACTORY, SOUNDLEVEL - 146 db ON SCALE A
GROUP 6 SHIELD	250	C–4	CENTER	SATISFACTORY 89% PROOF CHARGE
GROUP 6 SHIELD	280	C–4	CENTER	SATISFACTORY 100% PROOF CHARGE
GROUP 6 SHIELD	307	C–4	CENTER	SATISFACTORY 111% OVER PROOF CHARGE SHIELD RUPTURED
GROUP 6 SHIELD AND STORAGE BARRICADE	222	C–4	CENTER	SATISFACTORY

*	NOL	C-4
SERVICE CHARGE	567 gm	224 gm
PROOF CHARGE	708 gm	280 gm

Figure 8. Group 5 suppressive shield panels ready for assembly

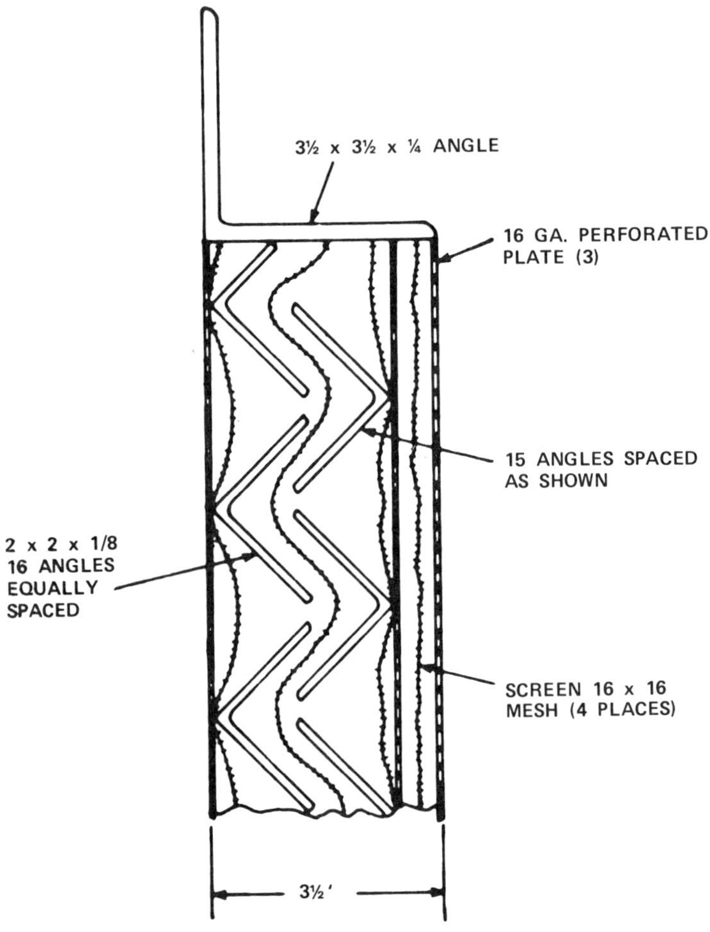

Figure 9. Group 5 panel section

Figure 10. Group 5 suppressive shield panels being assembled

Figure 11. Group 5 suppressive shield assembled, ready for test

into place. The modular characteristic simplifies the alteration of facilities to meet changing requirements.

Figure 11 is the assembled Group 5 Shield ready for test. The 55 gallon drum in the foreground gives a size reference. This shield has been approved for use with 1.84 pounds of C-4 or 30 pounds of illuminant mix.

The Group 5 Shield is especially designed for flame suppression. The structure is not as heavy as those intended to withstand the comparatively large transient blast overpressures of a detonation. The design has a large surface area and volume of steel to absorb heat and suppress flames. The venting of gaseous products of combustion precludes significant pressure and burning rate increases inside the shield with propellant or pyrotechnic material.

To determine limits on flame suppression capability of this shield, thermal suppression studies have been conducted beyond those required for the safety approval proof tests mentioned previously (5).

The tests with Shield Group 5 are summarized in Table II (4, 5).

Single base, multiperforated, M10 gun propellant in bulk was used in the propellant tests. The illuminant material was a 50:50 mix of sodium nitrate and powdered magnesium. The safety approval tests were conducted with a smaller charge weight than that shown, 30 pounds of the illuminant mix. To be complete, the proof test charge of 2.5 pounds is shown. That charge weight stressed the shield structure to its limit and was not increased.

Instrumentation layout for the tests is shown schematically in Figure 12. Instrumentation for the Group 5 Suppressive Shield tests is tabulated in Table III. Burning time was measured using photocells in the shield wall. Thermocouples in the bulk pyrotechnic were used to obtain an indication when the material was completely burned. Static overpressure was measured on large charges to estimate confinement effects. Radiant heat flux outside the shield was measured with Keithley 860 flux meters. Blast overpressure was recorded inside and outside the shield when explosive material was detonated in the shield. High speed motion picture coverage was included on all shots. Video display of each test in the instrument building was also recorded.

For comparison, free field temperatures and pressures were measured using instrument configuration shown in Figure 13. Thermocouples were on a line in one direction spaced at 5 foot intervals and pressure was measured in a direction perpendicular to the thermocouple line. Black and white, and color film coverage was also included.

Figure 14 is a photograph of a typical test set up. The open shield door was closed during the test.

Heat flux data transients five feet from outside of the shield (about 10 feet from the charge) are shown in Figure 15 with dashed lines. The solid lines are from heat flux

TABLE II. SUMMARY OF TESTS WITH SHIELD GROUP 5

MATERIAL	CHARGE WEIGHT	TYPE OF TEST
M10 GUN PROPELLANT	270 Kg (590 Lb.)	EXPLORATORY
ILLUMANT MATERIAL (50:50 SODIUM NITRATE–POWDERED MAGNESIUM)	22.7 Kg (50 Lb.)	EXPLORATORY
EXPLOSIVE MATERIAL (C-4)	1.1 Kg (2.5 Lb)	PROOF

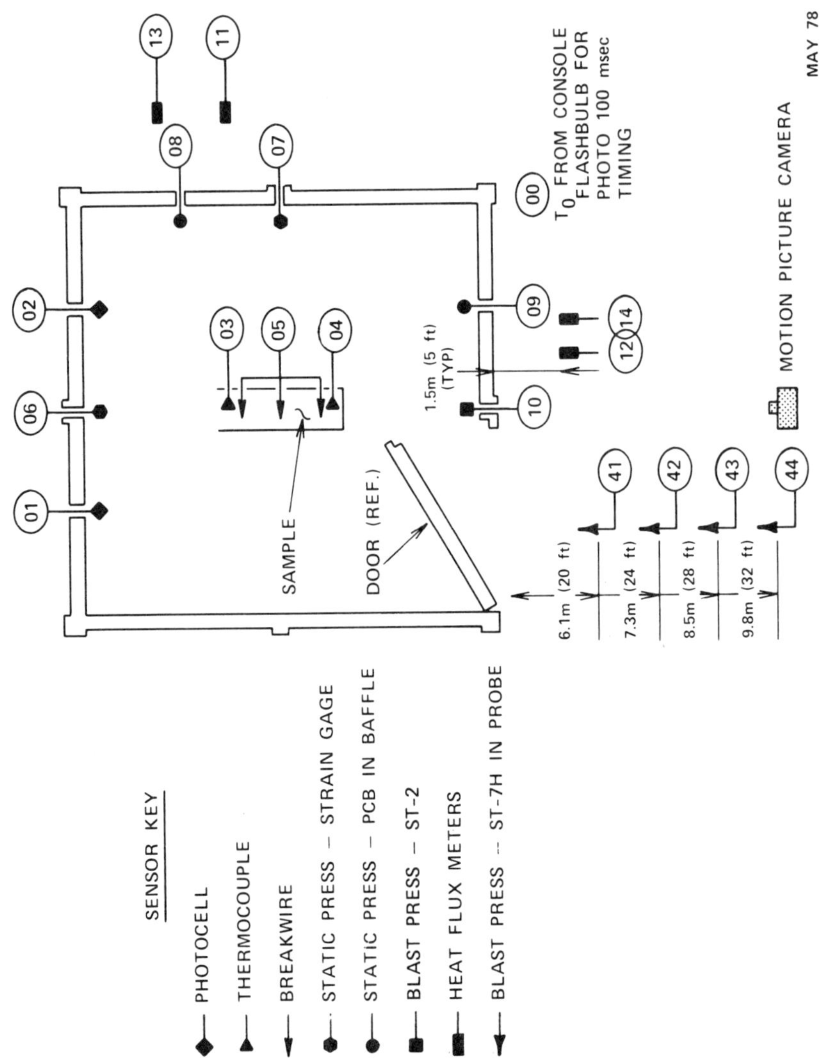

Figure 12. Sensor locations for Group 5 S/S tests

TABLE III

INSTRUMENTATION FOR GROUP 5 SUPPRESSIVE SHIELD TESTS

MEASUREMENT NUMBER	PARAMETER	TRANSDUCER	AMPLIFIER	INSTALLED TIME CONSTANT	RECORDER
00	TIMING	N/A	N/A	–	SANGAMO 4700
01	BURNING TIME	PHOTOCELL MONSANTO	TRANSDATA	1 msec	SANGAMO 4700
02	BURNING TIME	MT-2	NEFF109-6	1 msec	SANGAMO 4700
03 04	BURN RATE	FE-CONSTANTAN THERMOCOUPLE	NEFF109-6	100 msec	SANGAMO 4700
05	BURN RATE	BREAKWIRE	N/A	1 msec	SANGAMO 4700
06 07	STATIC PRESS.	MB151-DBZ-177 IN TUBE	NEFF109-6	10 msec	SANGAMO 4700
08 09	STATIC PRESS.	PCB101A02 IN BAFFLE MOUNT	NEFF109-6	10 msec	SANGAMO 4700
10	BLAST PRESS. (FACE-ON)	ST-2 IN WALL MOUNT	PCB401A13	200 msec	SANGAMO 4700
11 12 13 14	HEAT FLUX	KEITHLEY 860	N/A	1 sec	SANGAMO 4700
41 42 43 44	BLAST PRESS.	ST-7H IN AERODYNAMIC PROBE	PCB401A11	200 msec	BIOMATION 610B

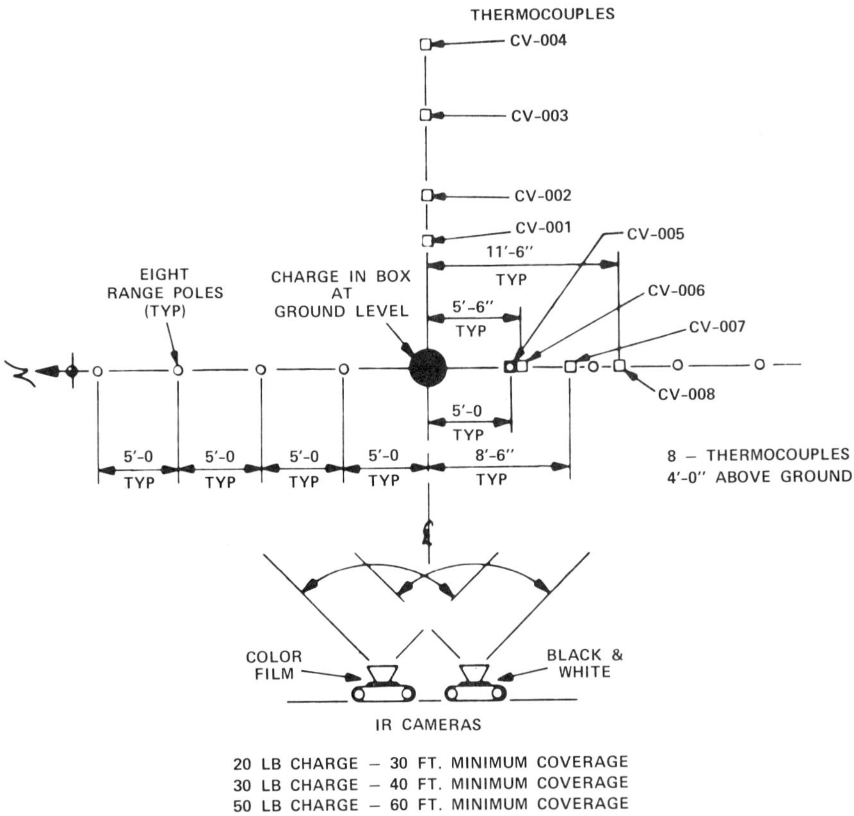

Figure 13. Free field illuminant test configuration

Figure 14. Shield Group 5 typical test set up

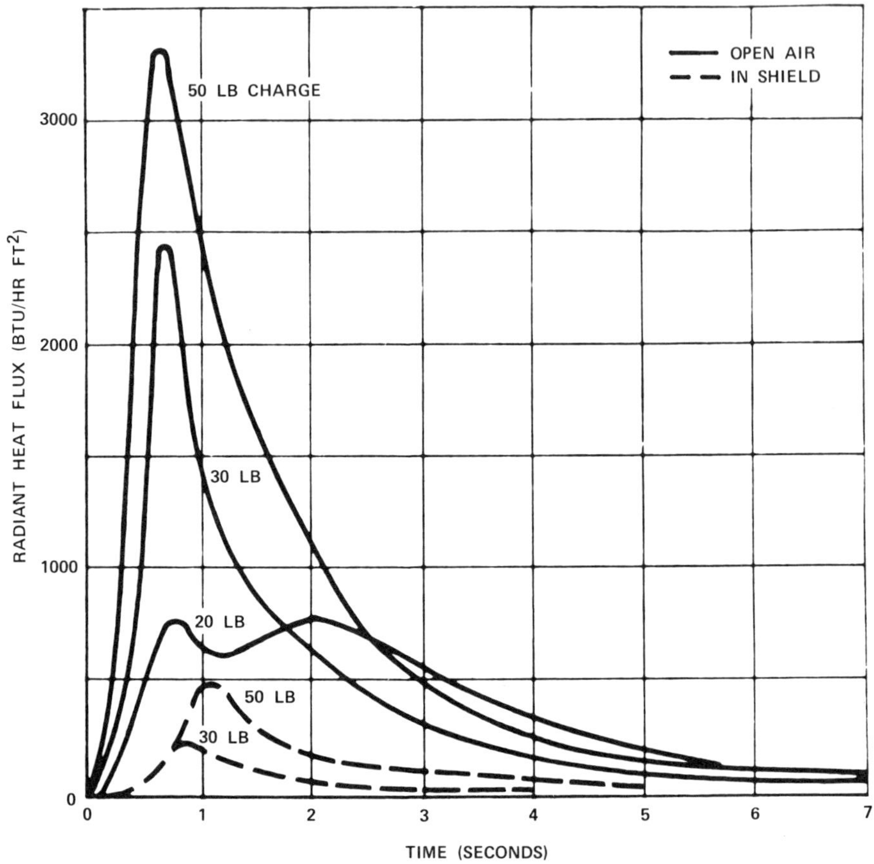

Figure 15. Heat flux as a function of time

TABLE IV. RESEARCH NEEDED

HAZARD DEFINITION

IMPROVED THERMAL SUPPRESSION TECHNIQUES

SUPPRESSED FIREBALL CHARACTERISTICS
- DIAMETER
- BURNING TIME
- RADIANT FLUX

EFFECT OF NON-UNIFORM VENTING

measurements about 10 feet from charges burned in the open. This is the same total distance from the charge as for the shield tests. Comparison between open air and shielded heat flux shows 85% reduction in peak radiant flux for a shielded 50 pound charge.

Preliminary propellant tests using a maximum charge weight of 590 pounds of M-10 single base multiperforated propellant in bulk with 0.0185 inch web resulted in no pressure rise in the shield. High radiant heat flux outside the shield wall indicated a need for improved thermal suppression. This work is not finished. More study is needed in the areas listed in Table IV.

An exhaustive search of the literature to identify hazards and improved thermal suppression techniques has revealed a need for more research in suppression techniques.

Methods do exist for estimating free field radiant flux, fireball diameter, and burning time for unconfined pyrotechnic material, but there is, at present, no method to compute attenuated thermal effects when a suppressive shield is used. Predictive models are needed (6).

Investigation of nonuniform venting has been initiated, but that work is not complete. Much work is required to develop the basic technology necessary to design optimal shields for flame suppression.

As a result of extensive investigation of blast and fragment effects, the technology for those hazards is well understood (2, 7-14). Predictive techniques for suppressive shield performance in attenuating blast and fragment hazards have been developed (1, 15-20). The following discussion of safety approved shields and technology summary briefly indicates the scope and result of the investigation.

Safety Approved Shields

To insure that Department of Defense safety offices approve site plans which incorporate suppressive shields, we have designed, fabricated, and proof tested several designs as listed in Table V.

The characteristics of the shields approved by the Department of Defense Explosives Safety Board are summarized in this table. They include sizes and charge weights typical to munitions manufacturing, but they can be scaled up or down in size or charge weight to meet special laboratory requirements.

As this table indicates, suppressive shields are approved for use in hazardous operations involving explosive charge weights up to the equivalent of 37 pounds 50/50 pentolite for Group 3 and 30 pounds of illuminant mix for Group 5.

Approved shield sizes range from the 2 foot diameter spherical steel shell of Group 6 to the $11\frac{1}{4}$ foot diameter cylindrical Group 3 shield.

The operator safe distance shown is the distance from the

TABLE V SAFETY-APPROVED SUPPRESSIVE SHIELDS

SHIELD TYPE	MATERIAL LIMIT	OPERATOR SAFE DISTANCE	SIZE
		FT	FT
GROUP 3	37 LB OF PENTOLITE	6.2	11.25 DIAMETER x 10 HEIGHT
GROUP 4	9 LB OF PENTOLITE	19	9.2 x 13.1 x 9.3 HEIGHT
GROUP 5	1.84 LB OF C-4 30 LB OF ILLUMINANT MIX	3.7 2	10.4 x 10.4 x 8.5 HEIGHT
GROUP 6	13.6 OZ OF PENTOLITE	1	2 DIAMETER
81 mm	TWO 81 mm ROUNDS 2.8 LB OF C-4	3	14 x 18.7 x 12.4 HEIGHT

exterior wall that an operator can be located and not be injured by blast overpressure or flame venting from the shield when detonation or deflagation occur within the shield (1).

Technology Summary

The suppressive shield designs that we have been discussing are based on a major four-year technology development effort by organizations listed in Table VI.

The roles they played in developing suppressive shield technology are indicated in the table.

The lead organization for suppressive shield technology development was Edgewood Arsenal, now called Chemical Systems Laboratory. Recent Army reorganization has reassigned responsibility for suppressive shielding to Large Caliber Weapons Systems Laboratory, ARRADCOM, at Dover, New Jersey.

Technology development has proceeded along the lines illustrated in the technology flow chart, Figure 16.

Hazards identified are classified as blast, fragment, and fireball. Description of each of these hazards is essential to design of a shield. Each of these hazards poses a special problem to the designer and requires consideration not only in terms of its own features, but also in terms of combined effects of all hazards acting together.

The next step is to develop procedures to predict suppression of blast, fragment, and flame hazards. The nature of the suppression governs the magnitude of the loads imposed on the structure. A safe, economical shield must be designed to withstand loads imposed. Suppression and design tradeoffs are made to obtain the best shield which satisfies hazard suppression requirements to provide a safe environment at minimal cost.

On blast environment, a predictive capability for characteristics of free air blast is available in the literature (2, 7). Techniques have been developed for defining internal transient and quasi-static blast overpressures (21, 22), pressure loads on the shield (2, 23, 24, 25, 26) and attenuated pressure external to the shield (1, 15, 16, 18, 20, 27).

The second major element to be considered in the design of a suppressive shield is the fragment threat. As we initiated our technology studies we found much was known about primary fragment hazards but little was known about secondary fragment hazards (8, 9, 12, 13, 28, 29). Primary fragments are those from material in direct contact with detonating composition. Secondary fragments are from surrounding equipment, not in direct contact with the composition that detonates.

It was necessary to establish a methodology to predict secondary fragment hazards and fragment suppression characteristics of the composite structural steel walls of a suppressive shield (19).

As a result of the Suppressive Shielding Program,

TABLE VI

SUPPRESSIVE SHIELDING

APPLIED TECHNOLOGY PARTICIPANTS

ORGANIZATION	ROLE
CHEMICAL SYSTEMS LABORATORY	TECHNICAL DIRECTION
BALLISTICS RESEARCH LABORATORY	MODELING – BLAST, FRAGMENTATION
NASA - NATIONAL SPACE TECHNOLOGY LABORATORY	FABRICATION AND TESTING
SOUTHWEST RESEARCH INSTITUTE	ANALYSIS – SCALING LAWS
NAVAL SURFACE WEAPONS CENTER	COMPUTER CODES – BLAST SUPPRESSION
U.S. ARMY CORPS OF ENGINEERS	DYNAMIC STRESS ANALYSIS – STRUCTURES
DUGWAY PROVING GROUND	FABRICATION AND TESTING
U.S. ARMY MATERIAL SYSTEMS ANALYSIS ACTIVITY	COST EFFECTIVENESS ANALYSIS
AAI CORPORATION	ENGINEERING SUPPORT

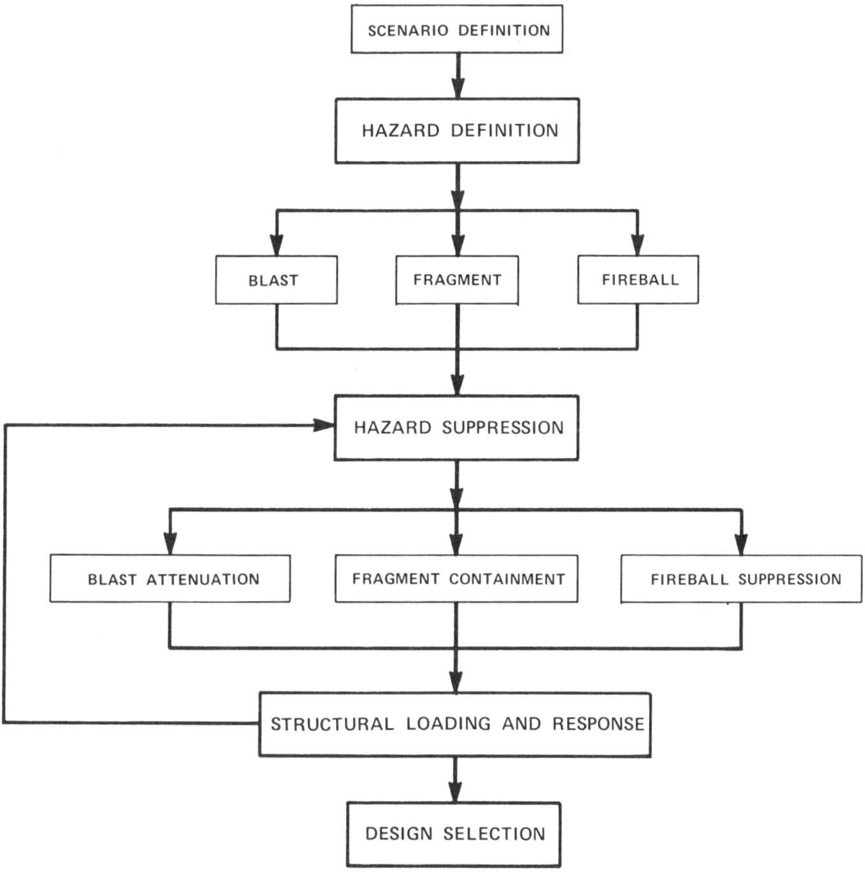

Figure 16. Technology flow chart

TABLE VII

SUPPRESSIVE SHIELDING ENGINEERING DESIGN HANDBOOK

CHAPTER I,	INTRODUCTION	
		– HISTORY OF THE PROJECT
		– USE OF HANDBOOK
CHAPTER II,	SAFETY APPROVED SUPPRESSIVE SHIELDS	
		– DESCRIPTION AND USE OF APPROVED SHIELDS
CHAPTER III,	EXPLOSIVE ENVIRONMENTS	
		– METHODS FOR COMPUTATION OF HAZARDS AND SUPPRESSION
CHAPTER IV,	SUPPRESSIVE SHIELD STRUCTURAL BEHAVIOR	
		– MATERIAL PROPERTIES AND ACCEPTABLE RESPONSE
CHAPTER V,	STRUCTURAL DESIGN AND ANALYSIS	
		– METHODS FOR DYNAMIC STRESS ANALYSIS
		– DESIGN PROCEDURES
CHAPTER VI,	STRUCTURAL DETAILS	
		– DOORS, PENETRATIONS, LINERS
CHAPTER VII,	ECONOMIC ANALYSIS	
		– METHODS
CHAPTER VIII,	ASSURING STRUCTURAL QUALITY	
		– QUALITY ASSURANCE
APPENDIX		
		– ENGINEERING DRAWINGS OF SAFETY APPROVED SUPPRESSIVE SHIELDS
		– ENGINEERING DESIGN CHARTS
		– SAMPLE ECONOMIC ANALYSIS

engineering methodology is available for modifying or scaling approved designs to meet specific munitions plant requirements. Where approved designs do not exist to meet certain requirements there is an engineering methodology for design and proof test of new shields.

This methodology is presented in an engineering design handbook for suppressive shields, published by the US Army Corps of Engineers, Huntsville Division, Huntsville, Alabama (30).

Table VII is a list of the chapter titles in the handbook. The information contained in each chapter is also shown.

Methods for modifying suppressive shields to meet specific production line requirements are given in Chapter II - Safety Approved Shields.

If a new shield must be designed, information in Chapters III, IV and V are used.

Chapter VI - Structural Details, has recommended designs for personnel doors, conveyor doors, as well as other penetrations for utilities and the like.

Economic analysis methods and quality assurance factors are included.

With this handbook, scientists or engineers requiring hazardous operation protection can select, modify, or design a suppressive shield for their required use. This handbook is available through DDC or National Technical Information Service and provides an alternative protective method previously not available to provide increased protection to personnel involved in hazardous operations.

Additional information on this subject is contained in references 31 - 90.

Abstract

A relatively new concept is presented for protection to the area surrounding hazardous work with explosives. Suppressive shields are vented composite steel structures which are designed to confine all fragments from an accidental detonation and to suppress hazardous blast and flame effects to a safe level.

Department of Defense Explosives Safety Board has approved five groups of suppressive shields for protection of munitions production operations in US Army Ammunition Plants. Safety approved shields encompass seven different designs which range in size from a two foot diameter steel shell (Shield Group 6) to a ten foot diameter steel cylinder (Shield Group 3).

Studies have been conducted to develop a technological base for accurate determination of shield performance parameters. It was found early in the program that the available data base was inadequate for accurately predicting the blast, fire, and fragment effects that would occur as a result of an accidental detonation of an explosive in a shield environment. Indepth studies resulted in development of suppressive shield design procedures published in the Suppressive Shield Design Handbook which provides information needed for design of facilities where explosive materials are used in hazardous operations.

Details on safety approved shields will be presented with technology study results.

Literature Cited

1. Katsanis, D.J., Safety Approval of Suppressive Shields, EM-TR-76088, Edgewood Arsenal, Aberdeen Proving Ground, Md., August 1976. (U)
2. Structures to Resist the Effects of Accidental Explosions, TM5-1300, Department of the Army Technical Manual, NAVFAC P-397, Department of the Navy Publication, AFM 88-22, Department of the Air Force Manual, Washington, D.C., June 1969. (U)
3. Nelson, K.P., Spherical Shields for the Containment of Explosions, EM-TR-76096, Edgewood Arsenal, Aberdeen Proving Ground, Md., March 1977. (U)
4. Koger, D.M. and McKown, G.L., Category 5 Suppressive Shield Test Report, EM-TR-76001, Edgewood Arsenal, Aberdeen Proving Ground, Md., October 1975. (U)
5. Wilcox, W.R., Shield Group 5 Suppressive Shield Plastic Liner and Propellant Testing, Contractor Report AR-TSD-CR-78001, NASA National Space Technology Laboratories, February 1978. (U)
6. Rakaczky, J.A., The Suppression of Thermal Hazards from Explosions of Munitions: A Literature Survey, BRL Interim Memorandum Report No. 377, Aberdeen Proving Ground, Md., May 1975. (U)
7. Engineering Design Handbook: Explosions in Air, Part I, AMC Pamphlet 706-181, Headquarters U.S. Army Materiel Command, Latest Edition. (U)
8. Headey, John, et al, Primary Fragment Characteristics and Impact Effects on Protective Barriers, PA TR 4903, Picatinny Arsenal, Dover, N.J., December 1975. (U)
9. Henry, I.G., The Gurney Formula and Related Approximations for the High Explosive Deployment of Fragments, PUB-189, Hughes Aircraft Co., AD813398, April 1967. (U)
10. Kingery, C.N., Air Blast Parameters Versus Distance for Hemispherical TNT Surface Bursts, BRL Report No. 1344, Aberdeen Proving Ground, Md., September 1966. (U)
11. Levmore, S., Air Blast Parameters and Other Characteristics of Nitroguanidine and Guanidine Nitrate, PA TR 4865, Picatinny Aresenal, Dover, N.J., November 1975. (U)
12. Mott, N.F., A Theoretical Formula for the Distribution of Weights of Fragments, AC 3742 (British) March 1943. (U)
13. Mott, N.F., The Theory of Fragmentation, AC 3348 (British) January 1943. (U)

14. Petes, J., "Blast and Fragmentation Characteristics," Annals of the New York Academy of Sciences, Vol. 152, Article 1, October 1968, pp. 283-317. (U)
15. Esparza, E.D., Baker, W.E., and Oldham, G.A., Blast Pressures Inside and Outside Suppressive Structures, EM-CR-76042, Report No. 8, Edgewood Arsenal, Aberdeen Proving Ground, Md., December 1975. (U)
16. Keenan, W.A. and Tancreto, J.E., Blast Environment from Fully and Partially Vented Explosions in Cubicles, Technical Report R828, Civil Engineering Laboratory, Naval Construction Battalion Center, Port Hueneme, Ca., November 1975. (U)
17. Oertel, F.H., Jr., Evaluation of Simple Models for the Attenuation of Shock Waves by Vented Plates, BRL Report No. 1906, Aberdeen Proving Ground, Md., August 1976. (U)
18. Proctor, J.F., Blast Suppression/Predictive Model, WBS 4333, Monthly Technical Report, November 1975, Naval Surface Weapons Center, White Oak, Silver Spring, Md. (U)
19. Ricchiazzi, A.J. and Barb, J.C., A Tentative Model for Predicting the Terminal Ballistic Effects of Blunt Fragments Against Single and Spaced Targets, A Comparison of Predicted and Experimental Results, BRL Memorandum Report 2578, Aberdeen Proving Ground, Md., January 1976. (U)
20. Schumacher, R.N. and Ewing, W.O., Jr., Blast Attenuation Outside Cubical Enclosures Made Up of Selected Suppressive Structures Panel Configurations, BRL Memorandum Report No. 2537, Aberdeen Proving Ground, Md., September 1975. (U)
21. Baker, W.E. and Oldham, G.A., Estimates of Blowdown of Quasi-Static Pressures in Vented Chambers, EM-CR-76029, Report No. 2, Edgewood Arsenal, Aberdeen Proving Ground, Md., November 1975. (U)
22. Kingery, C.N., Schumacher, R.N. and Ewing, W.O., Jr., Internal Pressures from Explosions in Suppressive Structures, BRL Interim Memorandum Report No. 403, Aberdeen Proving Ground, Md., June 1975. (U)
23. Gregory, F.H., Analysis of the Loading and Response of a Suppressive Shield When Subjected to an Internal Explosion, Minutes of the 17th Explosive Safety Seminar, Denver, Colorado, September 1976. (U)
24. Newmark, N.M., "Computation of Dynamic Structural Response in the Range Approaching Failure," Proceedings of the Symposium on Earthquake and Blast Effects on Structures, Los Angeles, Ca., 1952. (U)
25. Proctor, J.F. and Filler, W.S., A Computerized Technique for Blast Loads from Confined Explosions, 14th Annual Explosives Safety Seminar, New Orleans, Louisiana, 8-10 November 1972, pp. 99-124. (U)
26. Schumacher, R.N., Kingery, C.N., and Ewing, W.O., Jr., Air-blast and Structural Response Testing of a 1/4 Scale Category I Suppressive Shield, BRL Memorandum Report No. 2623, Aberdeen Proving Ground, Md., May 1976. (U)

27. Kinney, G.F. and Sewell, R.G.S., Venting of Explosions, Naval Weapons Center, NWC Technical Memorandum 2448, China Lake, California, July 1974. (U)
28. Hoggatt, C.R. and Recht, R.F., Fracture Behavior of Tubular Bombs, J. Appl. Physics, Vol. 39, No. 3, February 1968, pp. 1856-1862. (U)
29. Joint Munitions Effectiveness Manual, FM101-62-3, Manual of Fragmentation Data, 15 September 1973. (C-XGDS-3)
30. Suppressive Shields, Structural Design and Analysis Handbook, HNDM-1110-1-2, US Army Corps of Engineers, Huntsville Division, Huntsville, Alabama, 18 Nov 77. (U)
31. Ammunition and Explosives Ashore, Department of the Navy Publication, OP-5, including Volume 1, NAVSEA OP-5/1 "Safety Regulations for Handling, Storing, Renovation and Shipping", Volume II, NAVORD OP-5/2, "Storage Data", Volume III, NAVSEA OP-5/3, "Advance Bases", and all revisions thereto. (U)
32. Ammunition and Explosives Safety Standards, DOD 5154.4S, Department of Defense, Office of the Assistant Secretary of Defense (Installations and Logistics), March 1978. (U)
33. Baker, W.E., The Elastic Plastic Response to Thin Spherical Shells to Internal Blast Loading, Paper No. 59-A-95, Applied Mechanics Division, American Society of Mechanical Engineers, February 1959. (U)
34. Baker, W.E., et al, Analysis and Preliminary Design of a Suppressive Structure for a Melt Loading Operation, EM-CR-76056, Edgewood Arsenal, Aberdeen Proving Ground, Md., May 1976. (U)
35. Beedle, L.S., Plastic Design of Steel Frames, John Wiley and Sons, Inc., New York, N.Y., 1958. (U)
36. Biggs, J. M., Introduction to Structural Dynamics, McGraw-Hill Book Co., New York, N.Y., 1964. (U)
37. Building Code Requirements for Reinforced Concrete, ACI Standard 318-71, American Concrete Institute, Detroit, Michigan, 1971. (U)
38. Building Construction Cost Data, Robert Snow Means Co., Inc., 100 Construction Plaza, Duxbury, Mass., Latest Edition. (U)
39. Crawford, R.E., Higgins, C.J. and Bultmann, E.H., The Air Force Manual for Design and Analysis of Hardened Structures, AFWL TR 74-102, Air Force Weapons Laboratory, Kirtland AFB, N.M., October 1974. (U)
40. Dede, M., et al, Preliminary Estimate of Concrete Thicknesses and Construction Costs of Laced Reinforced Concrete Structures, Technical Report No. 4441, Picatinny Arsenal, Dover, N.J., October 1972. (U)
41. Design of Structures to Resist the Effects of Atomic Weapons, Structural Elements Subjected to Dynamic Loads, US Army Corps of Engineers Manual TM5-856-4, December 1965. (U)
42. Design of Structures to Resist Nuclear Weapons Effects, American Chemical Society of Civil Engineers Manual of Engineering Practice No. 42, ASCE, New York, N.Y. (U)

43. Explosive Safety Manual, AFM 127-100, Department of the Air Force, Washington, D.C., 2 December 1971. (U)
44. Final Report Application of Suppressive Structure Concepts to Chemical Agent Munition Demilitarization System (CAMDS), Report EA-FR-2B02, Edgewood Arsenal, Aberdeen Proving Ground, Md., 27 July 1973. (U)
45. Gray, B.H., Williamson, G.R., and Batson, G.B., Fibrous Concrete Construction Material for the Seventies, Construction Engineering Research Laboratory, Champaign, Ill., May 1972. (U)
46. Harris, C.M., and Crede, C.E., Shock and Vibration Handbook, McGraw-Hill Book Co., New York, N.Y., 1961. (U)
47. Healey, John, et al, Design of Steel Structures to Resist the Effects of HE Explosions, Technical Report 4837, Picatinny Arsenal, Dover, N.J., August 1975. (U)
48. Hubich, H.O. and Kachinski, R.L., Explosive Waste Removal Systems for Suppressive Shields, EM-CR-76002, Edgewood Arsenal, Aberdeen Proving Ground, Md., August 1975. (U)
49. Johnson, C. and Moseley, J.W., Preliminary Warhead Terminal Ballistic Handbook Part I, Terminal Ballistic Effects, NWL Report No. 1821, NAVWEPS Report No. 7673, US Naval Weapons Laboratory, Dahlgren, Virginia, March 1964. (U)
50. Kennedy, J.E., Explosive Output for Driving Metal, in Behavior and Utilization of Explosives in Engineering Design, ASME and Univ. of New Mexico, March 1972, Albuquerque, N.M. (U)
51. Kennedy, R.P., A Review of Procedures for the Analysis of Design of Concrete Structures to Resist Missile Impact Effects, Nuclear Engineering and Design, Vol. 37, 1976, pp. 183-203. (U)
52. Lasseigne, A.H., Static and Blast Pressure Investigation for the Chemical Agent Munition Demilitarization System: Sub-Scale, Rpt. EA-FR-4C04, November 30, 1973. (U)
53. Malvern, L.E., Propagation of Longitudinal Waves of Plastic Deformation in a Bar of Material Exhibiting a Strain-rate Effect, J. App. Mech., Vol. 18, pp. 203-8.
54. Manual of Steel Construction, Seventh Edition, American Institute of Steel Construction, Inc., New York, N.Y., 1970. (U)
55. Melin, J.W., Numerical Integration by Beta Method, ASCE Conference on Electronic Computation, Kansas City, Mo., 1968. (U)
56. Nagy, J., Et al, Explosibility of Miscellaneous Dusts, United States Department of the Interior, Bureau of Mines, Report of Investigations 7208, December 1968. (U)
57. Napadensky, H. and Swatosh, J., TNT Equivalency of Black Powder, IITRI TR J6289-4, IIT Research Institute, Chicago, Ill., September 1972. (U)
58. Napadensky, H. and Swatosh, J., TNT Equivalency of Large Charges of Black Powder, IITRI TR J6289-4, IIT Research

Institute, Chicago, Ill., February 1974. (U)
59. Napadensky, H., Swatosh, J., Humphreys, A., and Rindner, R., TNT Equivalency Three Pyrotechnic Composition, PA TR 4628, Picatinny Arsenal, Dover, N.J., June 1974. (U)
60. Nelson, K.P., The Economics of Applying Suppressive Shielding to the M483A1 Improved Conventional Munition LAP Facility, EM-TR-76087, Edgewood Arsenal, Aberdeen Proving Ground, Md., January 1977. (U)
61. Proctor, J.F., Internal Blast Damage Mechanisms Computer Program, 61 JTCG/ME-73-3, Joint Technical Coordinating Group for Munitions Effectiveness, April 1973. (U)
62. Roark, R.J. and Young, W.C., Formulas for Stress and Strain, McGraw-Hill Book Co., New York, N.Y., 5th Edition, 1975. (U)
63. Safety Criteria for Modernization and Expansion Projects, DRCPM-PBM Memorandum No. 385-3, Latest Edition. (U)
64. Safety Manual, AMCR 385-100, Hq, US Army Materiel Command, Alexandria, Va., Latest Edition. (U)
65. Salmon, C.G. and Johnson, J.E., Steel Structures-Design and Behavior, Intext Educational Publishers, Scranton, Pa., 1971. (U)
66. Seely, F.B. and Engisn, N.E., Analytical Mechanics for Engineers, John Wiley & Sons, Inc., New York, N.Y., 3rd Edition, 1948. (U)
67. Seely, F.B., and Smith, J.O., Advanced Mechanics of Materials, Second Edition, John Wiley and Sons, Inc., New York, N.Y., August 1961. (U)
68. Schroeder, F.J., Kachinski, R.L., Schnapfe, R.W., Koger, D.M., McKivrigan, J.L. and Jezek, B.W., Engineering Design Guidelines, Drawings, and Specifications for Support Engineering of Suppressive Shields, EM-CR-76097, Edgewood Arsenal, Aberdeen Proving Ground, Md., December 1976. (U)
69. Shear and Diagonal Tension, Report of ACI-ASCE Committee 326, ACI Manual of Concrete Practice-Part 2, American Concrete Institute, Detroit, Mich., 1968. (U)
70. Shelter Design and Analysis, TR-20 (Vol.4), Defense Civil Preparedness Agency, Washington, DC, July 1972.
71. Shields, Operational for Ammunition Operations, Criteria for Design of, and Tests for Acceptance, Mil Std 398, US Government Printing Office, Washington, DC, 5 November 1976. (U)
72. Study of Suppressive Structures Application to an 81 mm Automated Assembly Facility, Report EA 1002, Edgewood Arsenal, Aberdeen Proving Ground, Md., 16 April 1973. (U)
73. Swatosh, J., Cook, J., and Price, P., Blast Parameters of M26E1 Propellant, PA TR 4901, Picatinny Arsenal, Dover, N.J., December 1976. (U)
74. Swatosh, J., et al and Levmore, S., Blast Parameters of Lead Azide, and Tetracene, PA TR 4900, Picatinny Arsenal, Dover, N.J., December 1974. (U)

75. Swatosh, J. and Napadensky, H., TNT Equivalency N-5 Slurry and Paste, IITRI TR J6278, IIT Research Institute, Chicago, Ill., September 1972. (U)
76. Swatosh, J. and Napandensky, H., TNT Equivalency of Nitroglycerine, IITRI TR J6312, IIT Research Institute, Chicago, Ill., September 1973. (U)
77. Swatosh, J., et al, and Price, P., TNT Equivalency of M1 Propellant (Bulk), PA TR 4885, Picatinny Arsenal, Dover, N.J., December 1975. (U)
78. Swisdak, M.M., Jr., Explosion Effects and Properties Part I-Explosion Effects in Air, NSWC/WOL/TR 75-116, Naval Surface Weapons Center, White Oak, Silver Spring, Md., October 1975. (U)
79. Symonds, P.S., ASME Colloquium on Behavior of Materials Under Dynamic Loading, Ed., N.J. Huffington, November 1965. (U)
80. System Safety Program for Systems and Associated Subsystems and Equipment Requirements for, Mil Std 882, US Government Printing Office, Washington, D.C., 15 July 1969. (U)
81. Tanner, W.S. and Warnicke, C.H., Support Test for the Fabrication of a Suppressive Shield for a Naval Ordnance Disposal Facility, DPG-DR-74-304, US Army Dugway Proving Ground, Dugway, Utah, December 1973. (U)
82. Tomlinson, W.R., Jr., and Sheffield, O.E., Engineering Design Handbook, Properties of Explosives of Military Interest, AMC Pamphlet No. 706-177, Headquarters, US Army Materiel Command, January 1971. (U)
83. Trott, B. Dale, et al, Design of Explosion Blast Containment Vessels for Explosive Ordnance Disposal Units, Picatinny Arsenal, Dover, N.J., June 1975. (U)
84. Trott, B. Dale, et al, The Blast and Fragment Containment Capability of Portable Chambers, N00174-74-C-0219, Naval Explosive Ordnance Disposal Facility, Indian Head, Md., September 1975. (U)
85. Warnicke, C.H., Added Support Tests of the Suppressive Shield for Naval Ordnance Disposal Facility, DPG-DRI-74-313, US Army Dugway Proving Ground, Dugway, Utah, September 1974. (U)
86. Weibull, H.R.W., Pressures Recorded in Partially Closed Chambers at Explosion of TNT Charges, Annals of the New York Academy of Sciences, 152, Art. 1, pp. 356-361, October 1968. (U)
87. Wenzel, A.B., et al, An Economic Analysis of the Use of Suppressive Structures in the Lone Star Army Ammunition Plant 105-mm High Explosive Melt-Pour Facility, EM-CR-76032, Edgewood Arsenal, Aberdeen Proving Ground, Md., November 1975. (U)
88. Winter, G., et al, Design of Concrete Structures, McGraw-Hill Book Co., New York, N.Y., 1964. (U)

89. Zilliacus, S., Phyillaier, W.E. and Shorrow, P.K., The Response of Clamped Circular Plates to Confined Explosive Loadings, Naval Ship R&D Center Report 3987, NSRDC, Bethesda, Md., February 1974. (U)
90. 81 mm Suppressive Shielding Technical Data Package, Report EA-4E33 Edgewood Arsenal, Aberdeen Proving Ground, Md., January 1974. (U)

RECEIVED November 22, 1978.

3

Newly Developed Technology for Ecological Demilitarization of Munitions

F. H. CRIST

Ammunition Equipment Office, Tooele Army Depot, Tooele, UT 84074

The Army, in its role as storekeeper of munitions for all DoD services, has a continuing requirement to demilitarize unserviceable or obsolete munitions. Historically, demilitarization of munitions was accomplished by such expedients as sea dump or open air destruction. In 1970, the President signed Executive Order 11507 (later superseded by Executive Order 11752 dated 17 December 1973) directing that Federal Agencies set the example in abating pollution of the environment. This paper addresses some of the engineering efforts being expended to develop ecologically clean demilitarization technology that is safe for personnel handling this dangerous commodity. The candidate technology must also be affordable within the austere funding available for this important, but nevertheless, lower priority program that contributes relatively little to our defense posture. The demilitarization workload includes small arms ammunition, small to large caliber artillery ammunition, mines, mortar ammunition, rockets, bombs and myriad quantities of components that are used in the assembly of conventional munitions. Chemical munitions have received more intensified engineering and scientific effort to insure absolute safety of operators and positive retention of effluents generated by disposal operations.

Figure 1 shows a cross section of a rotary kiln type deactivation furnace developed to demilitarize small arms ammunition and various munition components. This equipment was designed to either burn or detonate the energetic material as it is moved by the helix flight from the cool feed end through the constantly increasing temperature of the 20-foot long retort. The deactivation furnace has proven to be an extremely cost effective system with all or much of the operating expense defrayed by the salvage value of decontaminated metals recovered by the process. Unfortunately, the high process feed rate generated large amounts of pollution to the atmosphere. A very high priority was therefore assigned to the development of emission controls for this process.

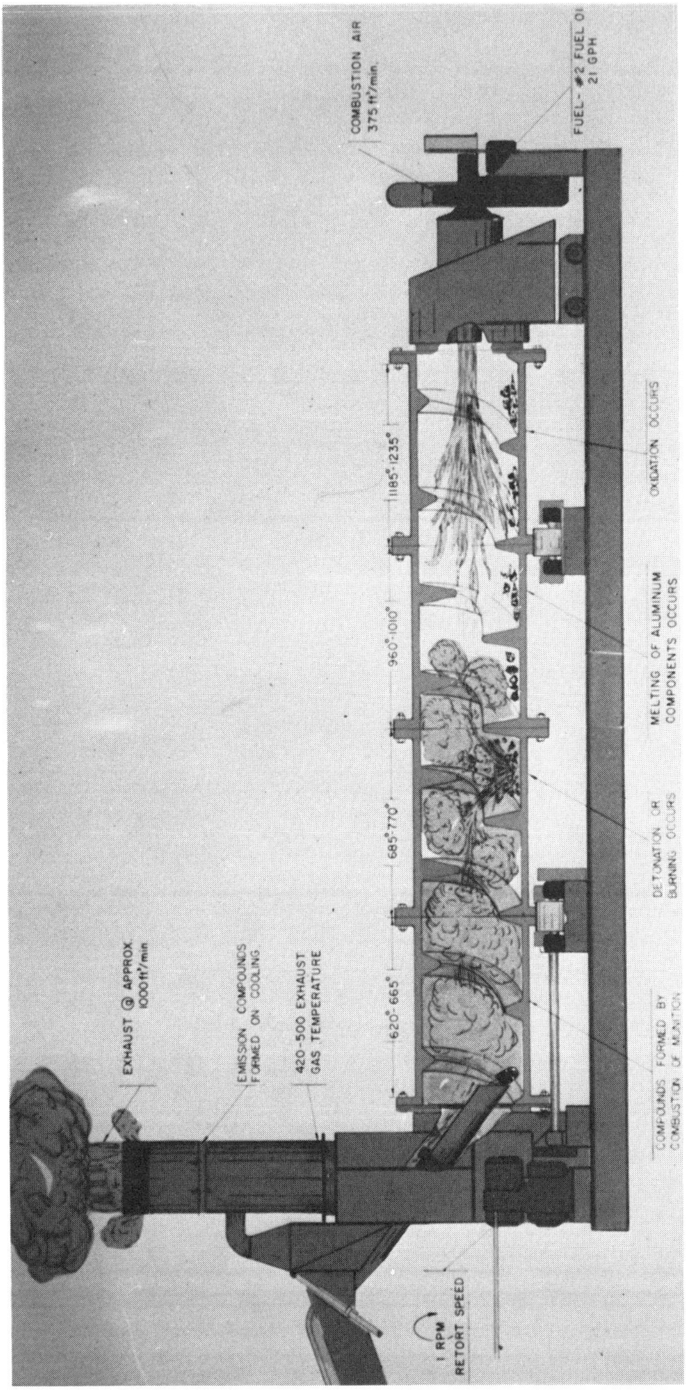

Figure 1. Furnace deactivation

Early in the emission control development program, it was recognized that the burning of myriad types of munitions and munition components could generate a horrendous number of chemical compounds in the effluent produced by the process. A thermochemistry computer program, developed at Edwards Air Force Base to describe the burning of rocket motors, was modified and implemented to predict products to be found in the effluent when burning different munition items. This program, through constant improvement and correlation with actual stack sampling during operations, proved to be extremely beneficial in optimizing the furnace operation to minimize pollutants of concern. After nearly one year of test burning with stack sampling accomplished by both private contractors and Government personnel, it was concluded that only particulate emission control was required to fully comply with all EPA and State standards. The control equipment selected consisted of a cyclone separator intended to remove large combustible particulates and sparks, spark arrestor screen and a standard bag house for removal of fine particles. The initial success of this system was marred by infrequent bag house fires. An intensive engineering investigation of these fires disclosed that the instrumentation used in prefatory and developmental testing was incapable of recognizing and responding to short duration, high temperature, exhaust gas excursions. The system was modified to permit a greater quantity of dilution air and a control system more responsive to short duration, high temperature surges. This modified system has proven to be extremely effective in abating pollution of the environment with no fires in the bag house. Most noteworthy is the fact that no additional operators were required and only one 25 hp motor was added to the total system consumption of energy. The system was designed so that additional control such as NO_x can be provided if required in the future.

A program to expand the application of the deactivation furnace for items with greater amounts of energetic material was undertaken simultaneously with the development of emission control equipment. Army engineers postulated and proved by tests that munitions could be degraded to expose energetic materials so that these materials would burn rather than detonate when introduced into the deactivation furnace. Figure 2 depicts live, high explosive, hand grenades degraded by the application of a small vent punched into its sidewall and munitions inerted by punching and processing them through the deactivation furnace. Munitions containing in excess of 1/2 pound of high explosive have been successfully sheared and burned in the deactivation furnace. Avoiding the cost of reversing the manufacturing process by recovering salvage metals that would have been lost in the open air destruction of these items provide major monetary benefits of this new technology. Shearing equipment, capable of matching the grenade burning rate of 22 rounds per minute, is now being completed to permit full scale production

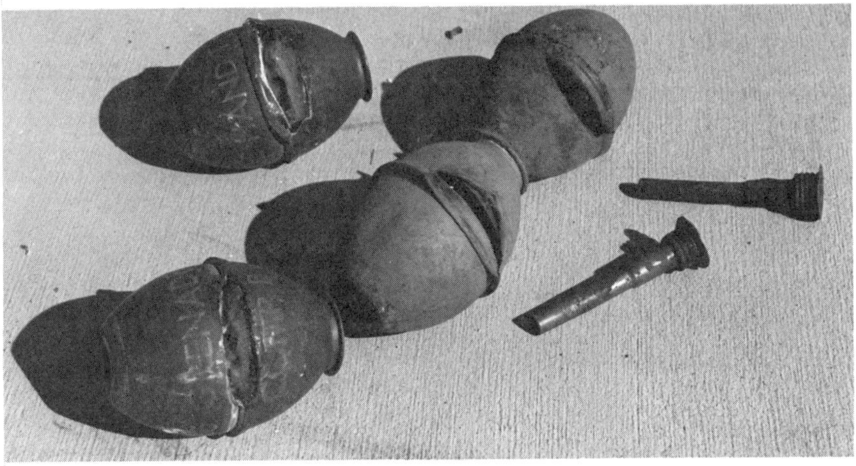

Figure 2. M26 grenades, sheared and burned

testing and operation.

Some time back, the Ammunition Equipment Office was tasked to develop a safe and ecologically clean demilitarization system for unserviceable nerve gas filled chemical munitions. Following the time-proven safety concept of having the least quantity of explosive filled munitions and the least number of operators present at the operation for the shortest possible time, Ammunition Equipment Office designed and developed the system shown at Figure 3. The machine developed to demilitarize the munitions is housed in a very substantial explosive containment vessel. This vessel was designed and has proven by dynamic test suitable for containing all fragments, gases and overpressures produced by a munition function that could occur during the demilitarization process. The equipment is designed to be monitored and totally controlled by computer. A manual override of some functions facilitated by closed circuit TV has also been provided. The only entry by personnel into the explosive containment cubicle will be for maintenance of the system. Work done within the explosive containment cubicle is shown in Figure 4. An M55 rocket contained in its shipping container is received on the inside of the explosive containment cubicle. All openings of the cubicle are closed, sealed and locked before the rocket is punched and drained of liquid nerve agent. After removal and transfer of the nerve agent by appropriate piping, the rocket, still in its container, is clamped in the machine, submerged under decon solution and severed into seven pieces by saws with six special circular blades. Major advantages of this technology development are the avoidance of all unpacking and disassembly operations, the elimination of personnel from the immediate demilitarization area and the relatively short (approximately four minutes) cycle time. The severed sections produced by this system are individually fed into the deactivation furnace where burning of the energetic material and intense heating by the burner within the retort guarantee total decontamination of the residue.

The technology developed for demilitarization of nerve agent filled rocket munitions has also been exploited in the demilitarization of conventional artillery projectiles. Projectiles are severed in suitable length pieces to permit burning of the unconfined energetic material in the deactivation furnace. More importantly, Ammunition Equipment Office engineers have postulated that 5 psi steam heating of the exterior casing of sectioned projectiles would melt the interface material between the steel sidewall and the explosive charge to permit the explosive charge to slip from the casing. Figure 5 depicts sample TNT and Composition B projectiles that were severed and steam heated for removal of the explosive charges. Explosive charges thus removed have been analyzed and found to be entirely suitable for recycling in the manufacture of new munitions.

An explosive washout process was developed several years

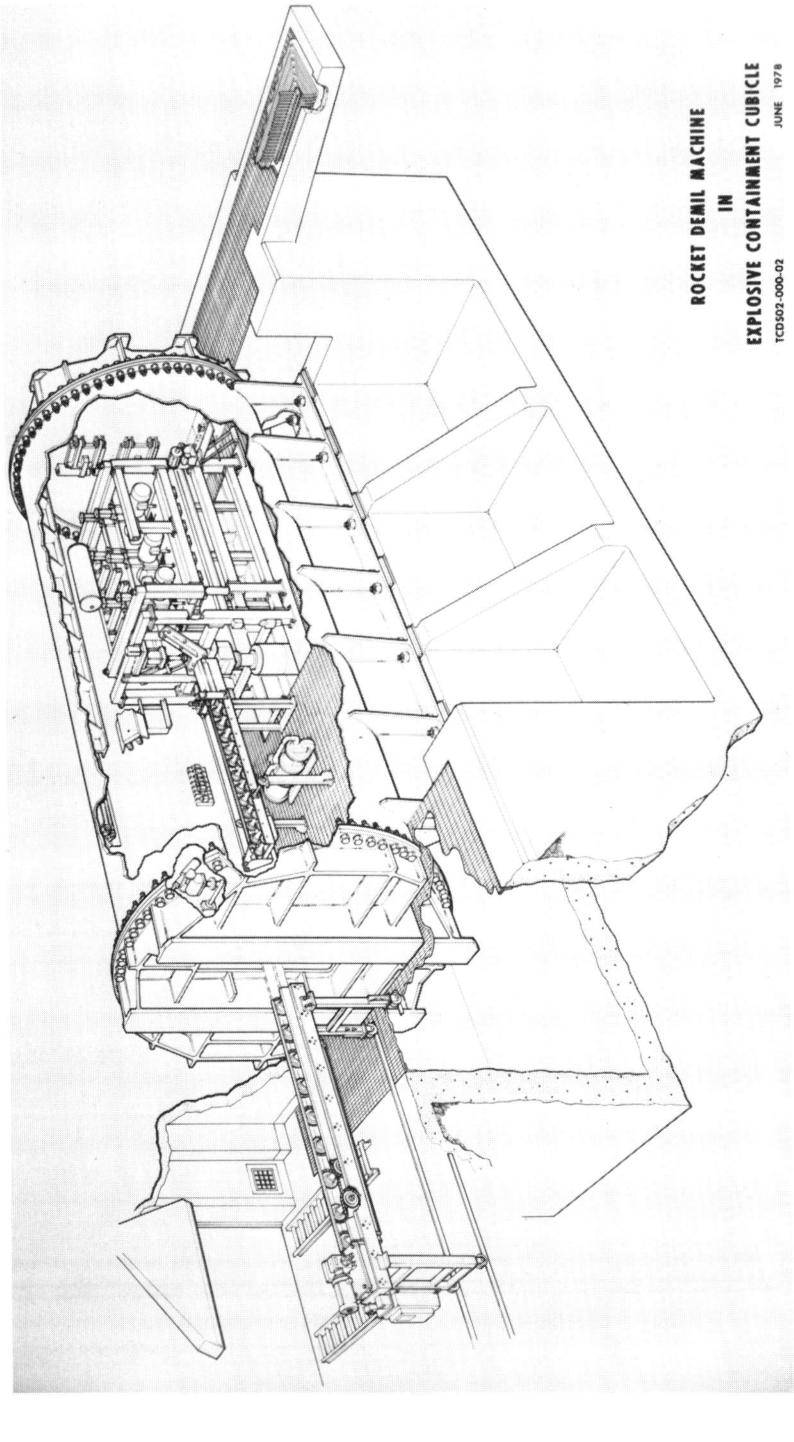

Figure 3. Rocket demil machine in explosive containment cubicle

Figure 4. M55 rocket before and after sectioning

Figure 5. TNT and composition B severed projectiles before and after melt out

ago for demilitarization of large caliber artillery projectiles and bombs that could not be open air detonated at depots in close proximity to population centers. In the washout process, hot water at high pressure melted and hydraulically eroded binary explosive from the interior of large caliber projectiles and bombs. The explosive was then recovered through a pelleting process for sale as a blasting agent. Contamination picked up in the washout process precluded recycling of the explosive for further military use. Treatment facilities were also provided to optimize the recycling of waters used in the process. Unfortunately, the system uses considerable energy for the accomplishment of its purpose. Recent, more stringent, regulation of the quality of liquid waste effluent would mandate a considerable investment to upgrade the pollution control equipment. A new method for removal of explosive from bombs and large caliber projectiles was sorely needed. The use of microwave energy was one of the candidate systems evaluated by Ammunition Equipment Office engineers. A small 5 kw microwave unit was rented and used for a series of melting tests. The test proved the validity of the system with two major bonuses: (1) Less than 20 kwhr of electric energy was consumed in the removal of explosives from a 750 pound bomb as compared to over 975 kwhr of equipment energy consumed by the washout process and (2) no liquid or gaseous effluent were produced by the microwave melt out of the explosive from the bombs. A production facility that will correct some problems encountered during testing and optimize the satisfactory test results is now being developed. Figure 6 depicts the ultimate facility visualized by the Ammunition Equipment Office engineers.

It has been common practice over the years to decontaminate downloaded ammunition parts or unserviceable machinery potentially contaminated with explosives by open air burning. Both the Clean Air Act and Subtitle C under the Resource Recovery Act preclude continuation of this procedure. A specially designed flashing furnace was procured and has been very successfully tested for the accomplishment of this work. The furnace was designed with workload capacity adequate to handle the products of several different demilitarization methods. The effluents generated by operation of this equipment are very similar to those generated by the deactivation furnace, therefore, it can share utilization of the same pollution control system. When operation of both the flashing furnace and the deactivation furnace is required, flashing furnace operation can be performed during an off shift if necessary.

White phosphorus filled munitions are a major item in the current Army demilitarization inventory. Ammunition Equipment Office engineers are developing methods to punch a sized hole into the white phosphorus filled munition to control the rate at which burning occurs when the munition is processed through the deactivation furnace. By controlling the burning rate, the

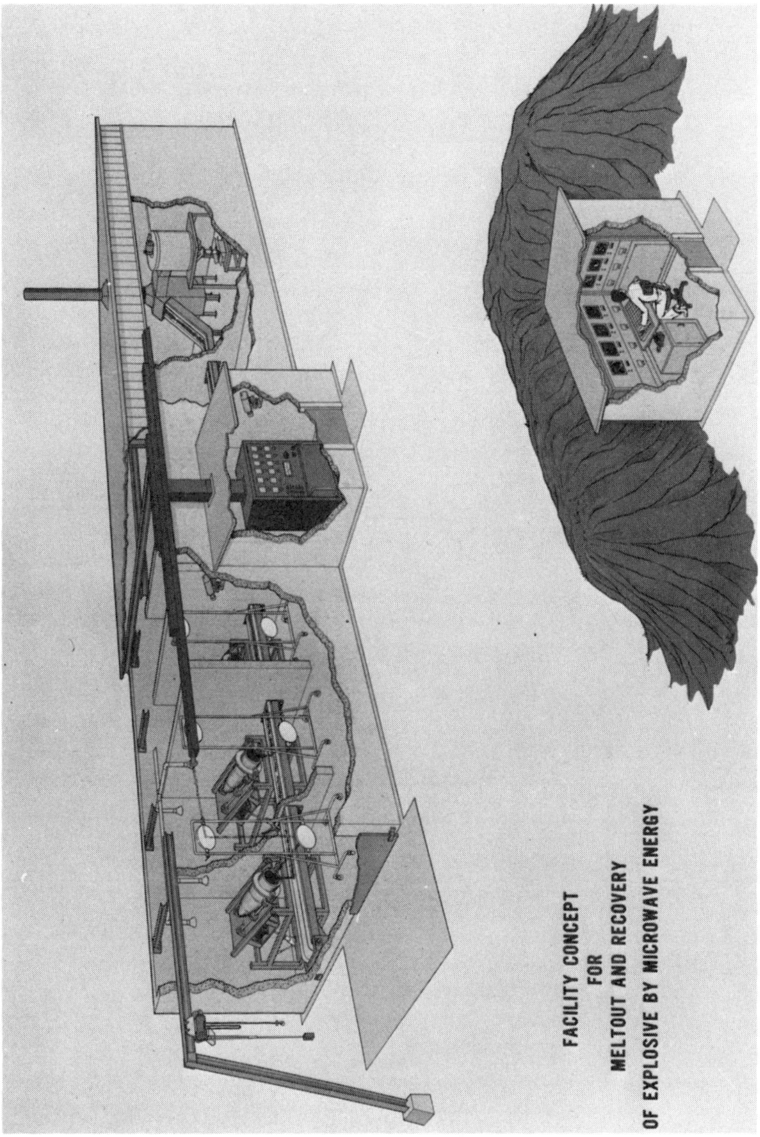

Figure 6. Facility concept for meltout and recovery of explosive by microwave energy

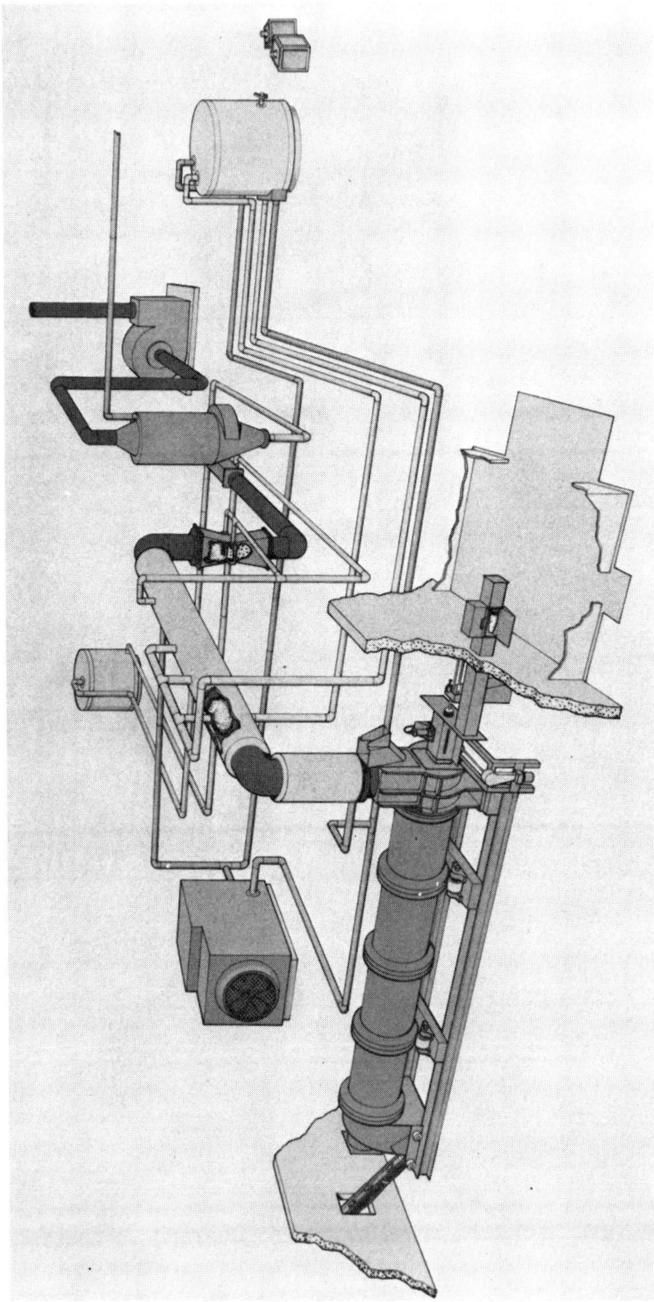

Figure 7. White phosphorus munitions disposal plant

Figure 8. Industrial robot

effluent can be processed through a hydrator and venturi scrubber for the production of relatively low grade phosphoric acid. This process would totally eliminate the need for open air burning of this item and is expected to net a slight profit over operating expenses. Figure 7 shows the Ammunition Equipment Office engineer's concept for the facility.

The demilitarization of munitions introduces some unusual hazards to operators performing the various tasks. Considerable effort has been expended to separate the operator from the immediate vicinity where the munition is being processed. Figure 8 depicts an industrial robot configured by Ammunition Equipment Office engineers specifically for the accomplishment of various ammunition handling operations. The robot is computer controlled with four separate computer programs, each capable of executing 128 discrete steps. Since procurement of the robot, a system utilizing very sensitive proximity sensors has been developed to recognize the geometry of a projectile at the pickup point and determine whether it is oriented correctly for placement in a machine for disassembly. If orientation is incorrect, binary signals from the proximity sensors trigger a subroutine of the robot program to rotate the projectile after pickup to correct its relationship with the disassembly machine. Many different applications of the robot have proven it to be extremely versatile and reliable for the performance of repetitive operations. The robot not only has the capability of handling munitions to and from disassembly equipment, but also monitors and controls the equipment.

Most of the demilitarization workload accommodated by new processes discussed in this paper include munitions developed during the Korean conflict. Many newer munitions now in storage are more sophisticated and complex in their assembly. Development of safe, cost effective and ecologically clean demilitarization capability for these items will be a keen challenge to the military and civilian engineering and scientific community involved in this work.

RECEIVED November 22, 1978.

4

Lightning and the Hazards It Produces for Explosive Facilities

RODNEY B. BENT

Atlantic Scientific Corp., P.O. Box 3201, Indialantic, FL 32903

Lightning is a natural phenomenon which poses a potential hazard to people, structures, and equipment unless adequate protection is provided. The type of protection required is related to the nature and function of the facility. The decision making process involves a number of interrelated factors which should be considered when determining the need for protection.

A knowledge of the basic lightning process can lead to a much better understanding of these lightning protection techniques and the resulting level of protection. The design of satisfactory lightning protection systems can, therefore, only be achieved with a knowledge of the mechanism and characteristics of a lightning strike and the related problems that a steep voltage wavefront has on inadequate bonding and grounding.

Lightning induced line surges can also cause major damage to electrical or electronic systems. A considerable proportion of the damage caused by such surges can be eliminated with careful planning of protection equipment. These line surges can also cause extra bits in computer software, which may lead to false decision making by the computer.

The Lightning Process in a Cloud-to-Ground Discharge

A cloud-to-ground lightning discharge is made up of one or more intermittent partial discharges. The total discharge, whose time duration is of the order of 0.5 seconds, is called a flash; each component discharge, whose luminous phase is measured in tenths of milliseconds, is called a stroke. There are usually three or four strokes per flash, the strokes being separated by tens of milliseconds. Often lightning as observed

0-8412-0481-0/79/47-096-079$12.25/0

by the eye appears to flicker. In these cases the eye distinguishes the individual strokes which make up a flash. Each lightning stroke begins with a weakly luminous predischarge, the leader process, which propagates from cloud-to-ground and which is followed immediately by a very luminous return stroke which propagates from ground-to-cloud.

It has been found that the electrostatic field takes about 7 seconds to recover to its predischarge value after the occurrence of a lightning flash at a distance beyond 5 km, but when the flash is very near, the recovery time may be different due to the presence of space charge. In both cases, regeneration of the field takes place exponentially.

Stepped Leader. The usual cloud-go-ground discharge probably begins as a local discharge between the p-charge region in the cloud base and the N-charge region above it (Figure 1). This discharge frees electrons in the N-region previously immobilized by attachment to water or ice particles. The free electrons overrun the p-region, neutralizing its small positive charge, and then continue their trip toward ground, which takes about 20 msec. The vehicle for moving the negative charge to earth is the stepped leader which moves from cloud-to-ground in rapid luminous steps about 50 m long, as shown in Figure 1. Each leader step occurs in less than a microsecond, and the time between steps is about 50 μ sec.

Return Stroke. When the stepped leader is near ground, its relatively large negative charge induces large amounts of positive charge on the earth beneath it and especially on objects projecting above the earth's surface (Figure 2). Since opposite charges attract each other, the large positive charge attempts to join the large negative charge, and in doing so initiates upward-going discharges. One of these upward-going discharges contacts the downward-moving leader and thereby determines the lightning strike point. When the leader is attached to ground, negative charges at the bottom of the channel move violently to ground, causing large currents to flow at ground and causing the channel near ground to become very luminous. The channel luminosity propagates continuously up the channel and out the channel branches at a velocity somewhere between 1/2 and 1/10 the speed of light. The trip between ground and cloud takes about 100 μ sec. When the leader initially touches ground, electrons flow to ground from the channel base and as the return stroke moves upward, large numbers of electrons flow at

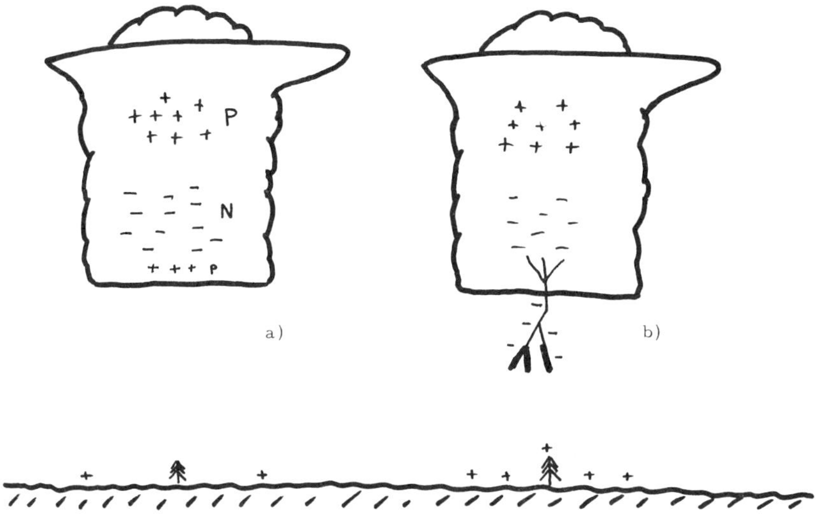

Figure 1. Stepped leader initiation. (a) Cloud charge prior to p-*N discharge, (b) stepped leader moving downward in 50-m steps.*

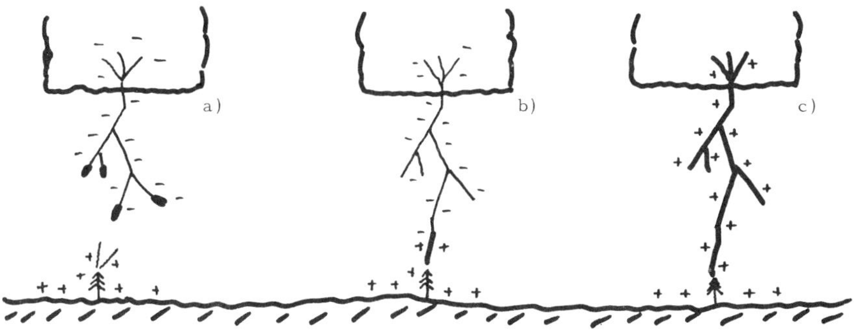

Figure 2. Return stroke initiation. (a) Start of upward moving sparks to meet leader; (b,c) return stroke propagation from ground to cloud.

greater and greater heights. Electrons at all points in the channel always move downward, even though the region of high current and high luminosity moves upward.

It is the return stroke that produces the bright visible channel. The eye is not fast enough to resolve the propagation of the return stroke, or the stepped leader preceding it, and it seems as if all points on the channel become bright simultaneously. A unique video photograph of the lightning leader with the return stroke part way up a double channel is shown in Figure 3.

After the first return stroke is complete, more charge may be made available to the top of the ionized channel and a dart leader will then pass down this branchless channel to ground, once more depositing negative charge. A second return stroke then passes up the channel. The process may continue several times in a fraction of a second.

Intracloud Discharge. Intracloud discharges have a duration of the order of 0.2 seconds, causing a continuous low luminosity in the cloud. It is thought that during this time a propagating leader bridges the gap between the two main charge centers. Superposed on the continuous luminosity are relatively bright luminous pulses which are probably relatively weak return strokes that occur when the propagating leader contacts a pocket of charge of opposite polarity to that of the leader.

Lightning Current. A full understanding of the time variation of a lightning current can only be obtained by oscillograph recording of the object struck. Because of the rare occurrence of lightning to an object of normal height, it is very unlikely that good statistics can be obtained without great expense and observations over a considerable period of time. Lightning currents, therefore, are mostly measured on tall structures or balloons connected to earth with steel cables. The currents from lightning have been found to be unidirectional and primarily discharge negative currents. Figure 4 shows a frequency-distribution curve of lightning current amplitudes, incorporating the best information available. The maximum value reliably recorded is 340k amps, but much higher amplitudes cannot be ruled out. The highest values occur primarily with the rare positive strokes.

A significant characteristic for lightning protection is the rate of rise of the current waveform. It has been generally assumed in the recent past that the lightning current waveform reaches a peak in some 1 to 3 μ sec. More recent information by

Figure 3. Photograph of lightning leader and return stroke

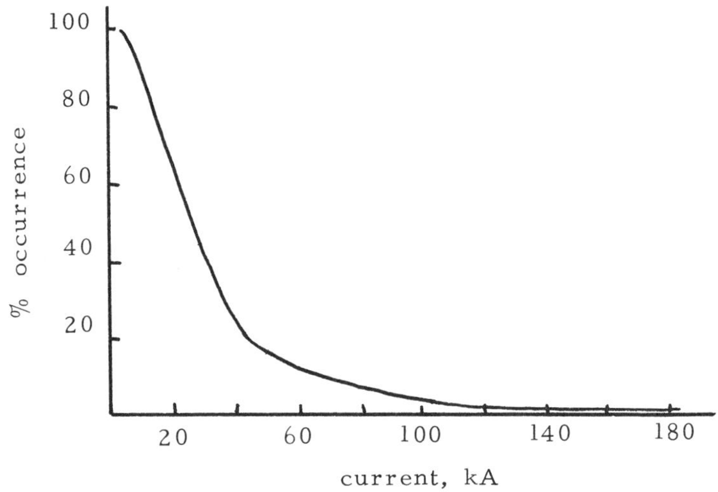

Figure 4. Distribution curve for lightning current amplitudes

Llewellyn (1) in 1977 indicates that the current risetime may be much less than 1 μ sec. Figure 5a shows some risetimes for a selection of close and distant storms. For some storms, the average time to peak value was found to be of the order of 1/4 μ sec. The effect of this term on lightning protection will be discussed later. Figure 5b shows normalized return stroke waveforms from over 2,000 discharges in several storms.

Breakdown Voltage. Breakdown tests of long spark gaps are made with impulse generators of 5 million volts or more where the test voltage rises at rates between 60 and 600 kV/μsec. Most of these tests are concerned with sparkover under a positive switching impulse because they produce much lower breakdown voltages than negative impulses. The minimum breakdown voltage is of concern to many engineers and Figure 6 gives representative test results by showing the variation of breakdown voltage for different electrode configurations and polarities. These tests were performed by Anderson and Tangen (2).

Further information on breakdown voltages are given in the book "Lightning Protection" by Golde (3).

Striking Distance. A normal negative leader progresses to earth in discrete steps until a counter streamer is initiated from ground or a grounded object. It is, therefore, at that particular distance that the point of strike is determined. This striking distance is defined as that distance between the tip of the lightning leader and the point to be struck. The main interest is limited to structures up to about 20 m height; structures of greater height begin to require special consideration as their height is greater than a leader step, and they may also induce upward going leaders. It is assumed after looking at Figure 6 that the critical breakdown between the tip of the lightning channel and earth is of the order of 5kV/cm for a normal negative stroke and 3kV/cm for the rare positive stroke. Small changes in this critical breakdown voltage gradient have little effect on the resulting value of the striking distance which is shown in Figure 7 and is taken from calculations by Golde (3).

The important conclusions from these results are that striking distance increases with the severity of the discharge; for an average 25k amp strike the distance is about 40 m. More important, however, these results and associated theory show that the progression of the leader remains quite unaffected by any feature on, or below ground, until the tip of the leader has reached a height of only a few tens or, at most, two hundred

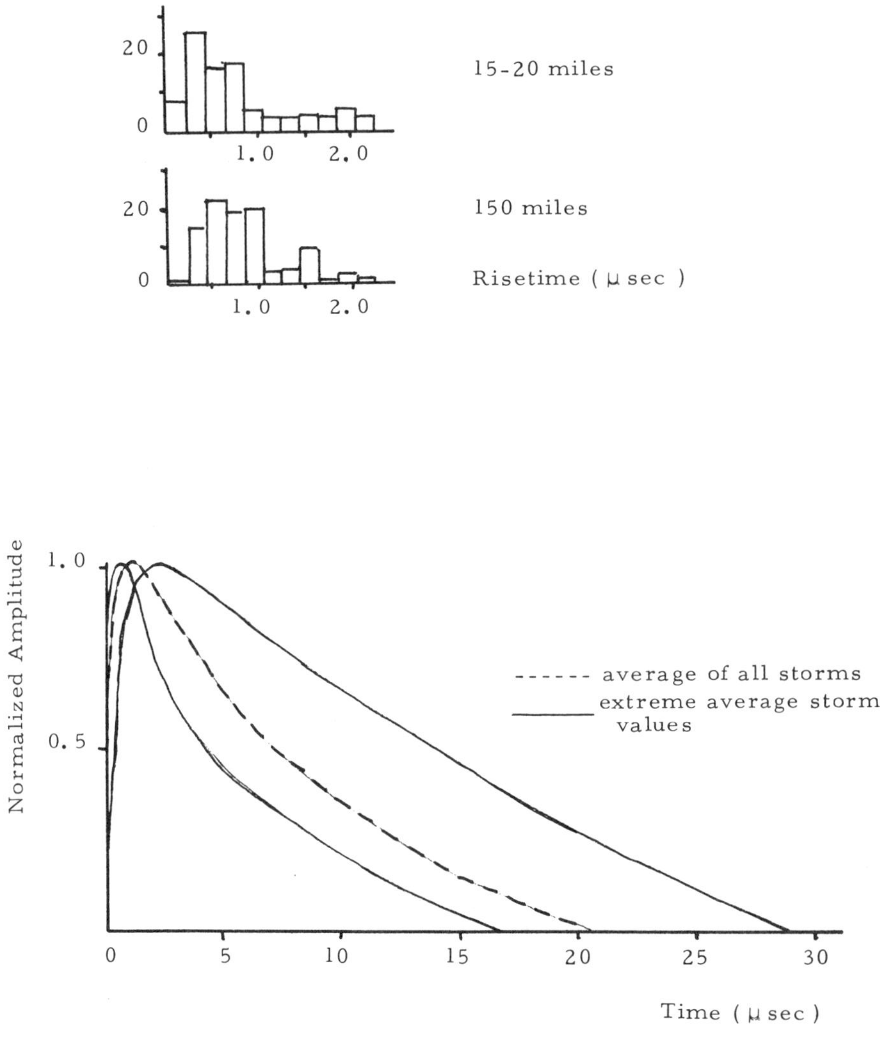

Figure 5. (Top) *Histogram of risetimes of return stroke magnetic waveforms for two storms;* (bottom) *normalized waveforms of the return stroke magnetic field for individual storms. (– – –) Average of all storms, (——) extreme average storm values.*

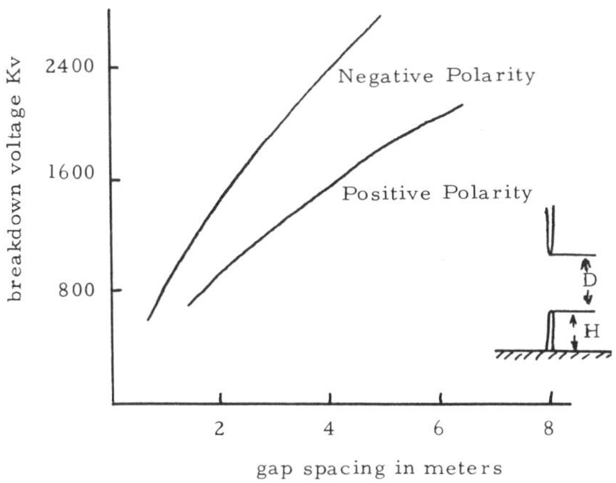

Figure 6. Switching impulse breakdown voltage for H/D = 1, which is average

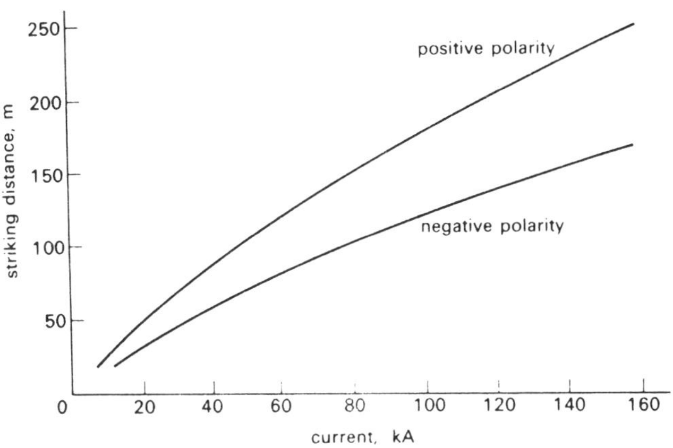

Figure 7. Variation of striking distance with current amplitude

meters above ground. These results, therefore, provide quantitative evidence against the belief in lightning concentration areas.

The Lightning Rod and Its Conductors

The Lightning Rod. It is a common misconception that lightning rods discharge clouds and thus prevent lightning. The rod only serves as a means to route the lightning harmlessly to ground by diverting the lightning when it approaches the striking distance discussed in the previous section. In the two hundred years since Benjamin Franklin investigated lightning, many manufacturers have tried to influence the public in the dissipation principle of lightning protection or elimination. This technique most certainly does not work and the lightning physicists' thoughts on this subject are discussed in masterly fashion by Golde in the following statement:

> "It is a manifestation of human weakness that a prejudice once acquired tends to be retained even in the face of overwhelming factual evidence contradicting the basis on which it was founded. In the realm of science a prejudice may be termed a misconception. Such a misconception which has persisted for over two hundred years and which is still widespread is the belief that a lightning conductor has the ability, or indeed the purpose, of dissipating silently the electric charge in a thundercloud thus preventing the "protected" building being struck".

There are several manufacturers of either radioactive lightning or lightning dissipation systems. The predominant scientific belief, however, is that neither of these systems are any benefit over the conventional lightning protection system. Extensive studies have been performed recently on dissipation arrays which only serve to enhance what scientists dating back to Franklin over 200 years ago have said; namely, that these arrays do no more than a conventional lightning rod and indeed probably do less. These studies have examined the historical, theoretical and experimental aspects of the arrays and have also investigated the arrays on site at several installations.

The arrays are believed by many non-scientists to give off significant corona discharge under thundercloud conditions so that the electrical characteristics of the storm are drastically changed. The claims indicate that a large area around the array is protected because this so-called excessive ionization either reaches the cloud and lowers it potential below the discharge level or provides a protective ion cloud over the protected area. These claims cannot be so as such a protective cloud would have to contain enough charge to make it more dangerous to the ground than the thundercloud itself. Secondly, corona ions dissipated from an array would recombine in normal air when only a few hundred feet above the ground, and there would also be less ions from an array than from a single conventional rod or even a few trees.

All factors related to these arrays indicate that they are not as good a lightning protector as a conventional single conductor. A U.S. Navy report (4) and FAA report (5) discuss both sides of the topic.

In order to examine the claims related to radioactive lightning rods, it is necessary to consider the physical process of a discharge to a conventional rod. When a lightning rod is in the area of a lightning leader the electric field around its tip would be extremely high and the air in this region would be in glow discharge which indicates millions of free electrons moving at the point. As the electric field increases, this ionization process or corona current also increases to arc discharge and a spark reaches out to meet the downward coming leader, forming a path for enormous currents to flow.

Radioactive rods contain a certain amount of radioactivity in the area near the tip of the rod, supposedly to enhance the ionization and hence attract the lightning leader over larger distances. These claims have been examined experimentally and theoretically by many scientists with negative results. In effect, the analysis shows the corona current from the radioactive rod is slightly higher than that from the conventional rod, as the manufacturer claims, only when electric fields are low such as under a fair sky. When a thunderhead approaches and the electric fields build up, however, the radioactive rod gives off less corona current than the conventional rod and is, therefore, less likely to be struck. This can be explained by the fact that the ionization cloud produced around the rod by the radioactive source provides an ion shield around the tip reducing its effectiveness in sending up the necessary upward leader spark.

The corona discharge from a conventional rod was found to exceed that from a radioactive rod by an order of magnitude under lightning-like electric fields indicating that radioactive rods are much less capable of influencing the path of a lightning discharge than a conventional rod. An example of failure of radioactive lightning rods was illustrated on the Vatican's Berni

Colonade building in Rome which is protected by such rods. On the 6th of March, 1976, the Papal Crest was struck by lightning and knocked off indicating failure of such a protective system.

The lightning rod has the purpose of intercepting a lightning strike and deflecting it from the structure. When several lightning rods are to be put on a building, one should develop a common-sense solution which will strike a reasonalble balance between protection and cost.

It is possible, with care, to use existing gutter and rain pipes to obtain protection at reduced costs, but care must be taken when incorporating modern metalic roofing materials to be part of the system. Lightning can penetrate metal sheets of 1mm thickness or more, and perhaps the cost of such repair might be acceptable. The minimum thickness is defined in certain codes to be 0.3mm for copper, and 0.5mm for other metals. Some roofs use metal foils of less thickness. A lightning strike to this type of roof will not only burn a large hole, but can cause large areas of foil to be torn off due to the mechanical effects.

Down Conductors. When lightning strikes an air terminal the injected current must be transferred by the shortest possible path to ground. The down conductor has this function, but because the inductance of this down conductor is a major factor in determining the occurrence of the dangerous side-flash to some internal grounded object, it must also have the lowest impedence that can be afforded.

The inductance of a down conductor is directly proportional to its height. By paralleling two down conductors their combined inductance is reduced to approximately one-half that of a single conductor and so on. The down conductors should not be spaced too close together however, otherwise the above rule is not accurate. The importance of having at least two down conductors is therefore a considerable advantage in reducing the dangerous side-flash, the action of which is discussed later. Right angle bends in a down conductor also increases the inductance and such a design needs careful consideration.

Once a lightning strike has been intercepted and passed to the surface of the earth, it is the function of earth electrodes to discharge the current into the ground. Two important factors are the ground resistance, which plays a part in side-flashing, and the potential distribution over the ground surface. If the ground resistivity is high, advantages can be achieved by bonding the down conductors to water pipes to lower the resistance to ground. The risk in side-flashing is thus determined exclusively by the inductance.

Side-flashing can also occur below the ground to buried metal pipes or wires and care must be taken in the design and positioning of the grounding electrodes. Typical values of impulse breakdown in soil are 2 to 5kV/cm, which leads to side-flashes of several meters. In air the value is 9kV/cm and brick and concrete has a slightly lower breakdown strength.

It is interesting to note that the length of a ground rod has a much more significant effect on the resistance than its radius. Curves demonstrating this effect are shown in Figure 8, which also implies that little benefit is achieved by extending the rod beyond 2 or 3 meters or increasing its diameter beyond 1.25 cm. Strip electrodes are beneficial where high resistivity ground exists below a layer of low resistivity.

Materials. The type of material used for roof and down conductors seems to be governed by tradition. Copper, aluminum and galvanized steel are all acceptable but there are conflicting opinions as to whether the material should be of rod, tube, strip or stranded form. Stranded copper is not deemed acceptable in the codes of several countries, although it is accepted in the USA code. Copper or copper alloys must not be used on a building with aluminum fittings.

A strong corrosive effect can be caused by rainwater dripping off copper conductors onto some metals such as zinc or lead which are often used on buildings. Dissimilar metals should be avoided as far as possible, and one should be aware that stranded materials are more severely attacked by corrosion than solid conductors.

Corrosion plays a high risk underground, in particular to aluminum which is totally unacceptable. The electrolytic properties of some soils cause corrosion to all these metals, as do stray currents produced by DC railway lines on DC high voltage systems where the earth is used as a return path. Cathodic protection can help eliminate this type of problem.

The Basic Requirements of Lightning Protection

The relative need for lightning protection at a facility is dependent on many factors as indicated by Smith (6):

(a) Type or usage of facility;
(b) Personnel safety;
(c) Prevalence of lightning;
(d) Type of construction;

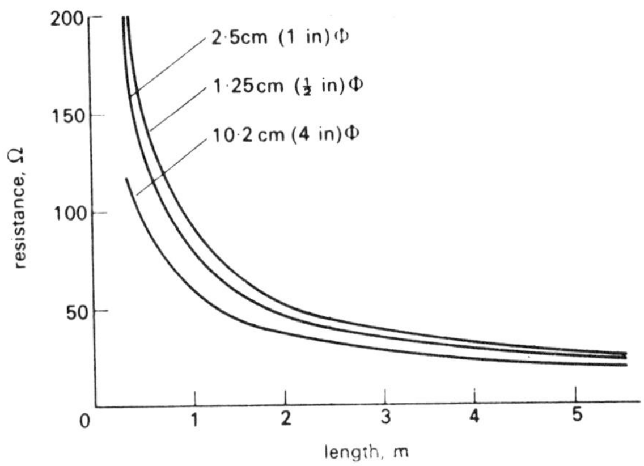

Figure 8. Variation of ground resistance of rod electrodes of different diameter with length (British code)

(e) Contents;
(f) Economic risks;
(g) Degree of isolation (relative height of surrounding structures);
(h) Type of terrain; and
(i) Height of structure.

The decision to provide protection may be based primarily on one factor alone. Some possible single factors can be personnel safety requirements, reduced insurance rates, or to meet an imposed lightning protection code.

Three basic requirements must be fulfilled to provide protection against direct lightning strikes to a structure:

(a) A conductive object must be provided to intentionally attract the leader stroke;
(b) A path must be established that joins this object to earth with such a low impedence that the discharge follows it in preference to any other; and
(c) A low resistance connection must be made with the body of the earth.

The lightning protection system should conform to an accepted set of guidelines. Codes and standards have been established in many countries to provide the necessary guidelines. Every code or requirement has one thing in common and that is diverting a direct strike to earth. The primary difference in the various codes is the philosophy used in achieving an effective protection system.

The United States has two nationally accepted codes; the National Fire Protection Association's Lightning Protection Code (ANSI C5. 1), and the Underwriter's Laboratories Master Labeled Lightning Protection System (Standard UL 96A). The requirements of these two codes are quite similar and are probably equally utilized on structures throughout the nation. The major difference between the two is that the Master Label can be certified upon both a factory inspection and labeling of the lightning protection materials and upon performance of a field inspection by an authorized inspector.

Lightning Protection by Overhead Wire

Danger facilities of large dimensions requiring the best possible protection should be provided with a system of catenaries suspended from tall masts. These catenary wires must pass to ground some distance away from the protected structure so that the lightning current may go to ground at a distant point.

The grounding wires should radiate away from the structure, be strip conductors and should not be bonded to the structure. The overhead wires should be far enough from the structure to eliminate side-flashing which was described earlier and the protective angles from the wires must cover the building structure.

Theoretical investigations of the electric field around elevated grounded structures are summarized in the following three figures. Figures 9 and 10 show the equipotential lines around an elevated grounded wire as drawn by a computer, and the field lines around a grounded wire at 250 feet during a thunderstorm. From Figure 10 it can be seen that if a weak lightning leader came toward ground within 80 feet of the wire it would be attracted to it. Hence, if two wires were spaced 160 feet apart any such leader approaching ground in that region would be attracted to the wires, thus protecting the structure under them. The protection area is a function of wire size and height, as shown in Figure 11.

Photographs taken of lightning striking a wire stretched over a canyon in New Mexico by Dr. Moore, verify that it did indeed strike the wire underneath on various occasions. At times when a strong lightning leader approaches, the function of the wire, then in a high corona discharge state, is to provide the upward leader thereby attracting the downward stroke to the wire.

If this type of lightning protection is carefully planned, chances of failure will be extremely small, however, the installation and maintenance problems of such a system may be considerable.

Electrical, Mechanical and Thermal Effects

Electrical Effects. No lightning strike to a structure has attracted more attention in the last decades than the so-called side-flash. It has been examined repeatedly and its dangers are illustrated in the technical literature. Its prevention must be provided in order to stop incidents in which a protected building has been struck and a person in such a building injured.

An illustration of the principles of the conditions leading to the risk of a side-flash are shown in a simple example in Figure 12.

The illustration shows the outlines of a building with a lightning conductor protecting the chimney which constitutes the highest point. In the attic is an electrical point or a water pipe which in turn is connected directly or indirectly to ground. Bear in mind that an electrical supply is still almost at ground poten-

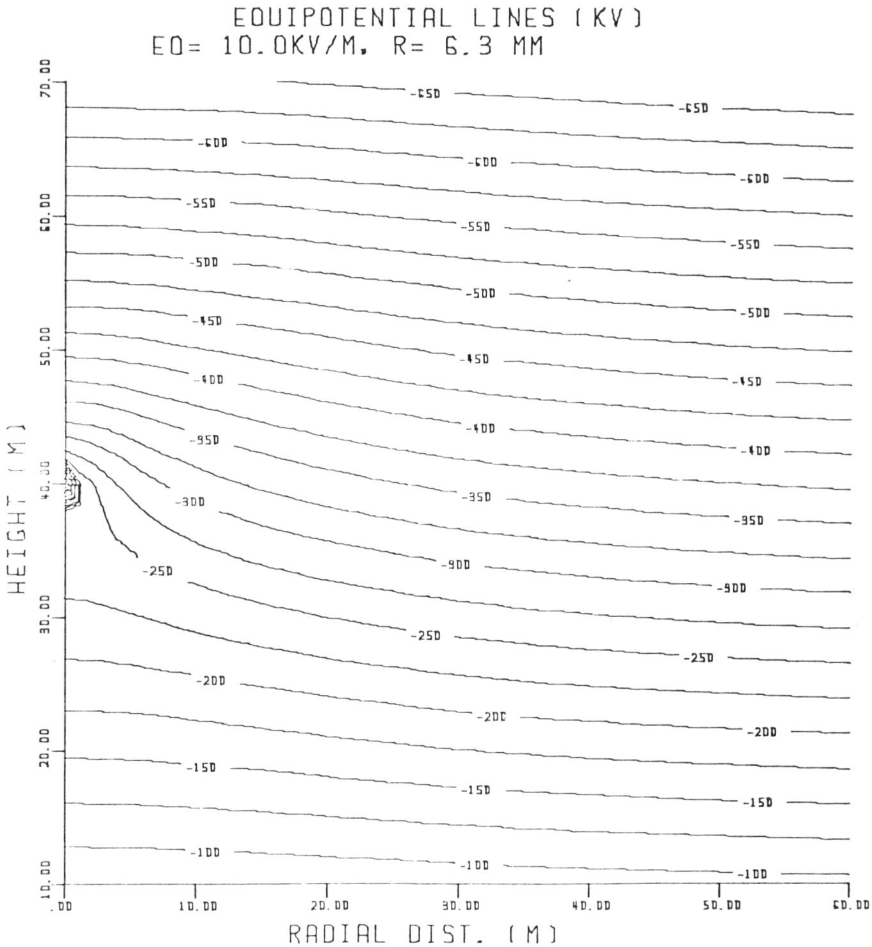

Figure 9. Equipotential lines around an elevated grounded wire

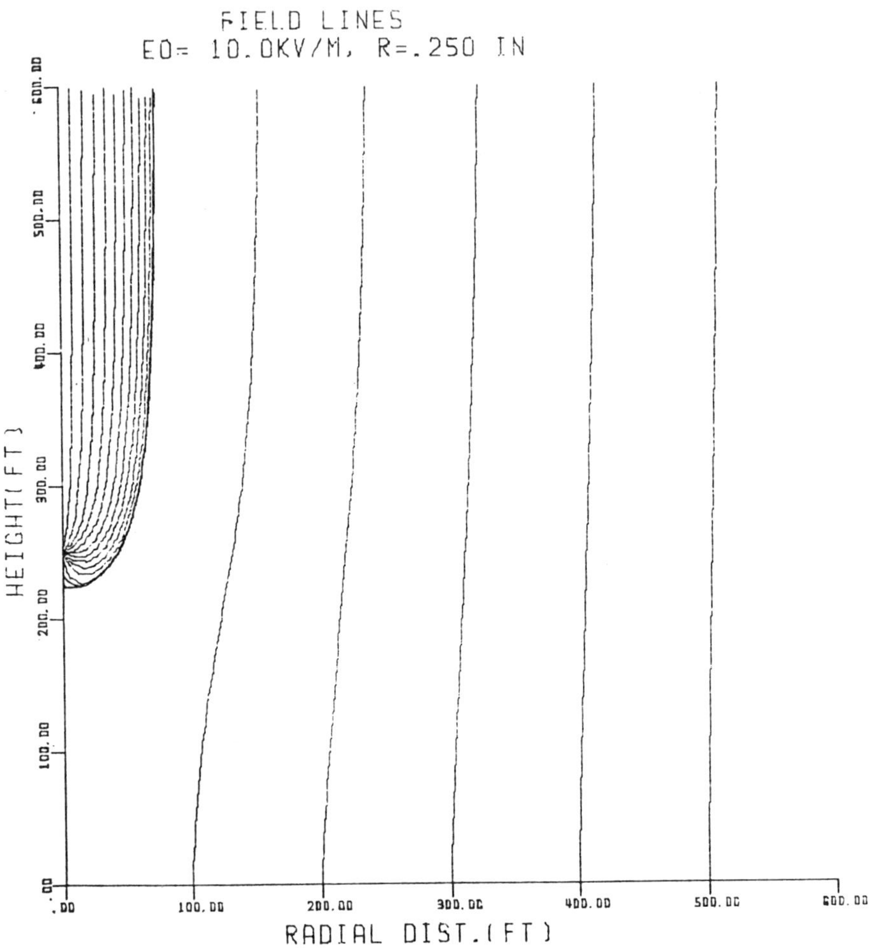

Figure 10. Electric field lines in the vicinity of an elevated grounded wire

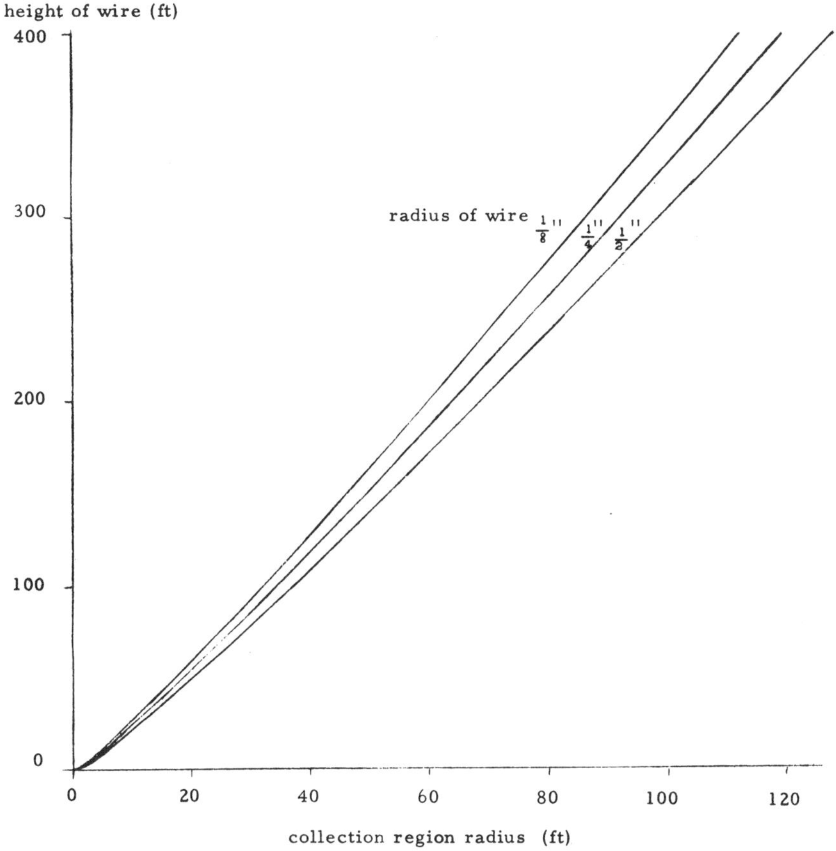

Figure 11. Field line collection area as a function of wire size and height

tial as far as the extremely high lightning voltages are concerned.

Let us now assume that the lightning conductor on the chimney is struck by a lightning current of amplitude i. The current is then discharged along the roof conductor, the single down conductor and into the earth electrode. This path constitutes an inductance L, while the impedance of the earth electrode may be described by its effective ground resistance R. The top of the lightning-protective system is thus raised to a potential with respect to true earth which is given by:

$$u = iR + Ldi/dt$$

For the purpose of a numerical estimate, we may assume an intense lightning current of crest value i = 100 k amp and a ground resistance of R= 10Ω. The inductance of a single vertical conductor is about 160 μH per 100 m and the rate of rise of the front of the lightning, as reported by Llewellyn (1), may be taken as 50k amp/$^1/_4$ μsec. If the height of the chimney above ground is 10 m, the top of the lightning conductor is raised to a potential with respect to true earth which amounts to:

$$u = 10^5 \times 10 + 10^{-1} \times 1.6 \times 10^{-4} \times 10^6 \times 2 \times 10^5 \text{ V}$$
$$= 10^6 + 3.2 \times 10^6 \text{ V} = 4.2 \text{ MV neglecting phase differences.}$$

In contrast, the internal grounded wire or pipe remains at ground potential even when the house is struck so that the potential difference of 4.2 MV is suddenly impressed between the lightning-conductor system and the internal point. If the electric breakdown strength of the clearance D is less than that potential difference, an electric breakdown occurs from the lightning conductor to the water tank; and this is termed a side-flash. The breakdown strength of air for a chopped impulse voltage can be taken as 900 kV/m, hence a 5m flash could materialize. Similar situations may occur if people are standing between a grounded air-terminal lead and a grounded instrument in the building. Side-flashes may also occur under the surface and in the process can throw up rocks and soil over great distances. The simple solution of bonding the down conductor to the grounded object will alleviate many problems.

Thermal Considerations. The lightning leader stroke has a narrow cone which is surrounded by much larger corona. The return stroke current is concentrated in this central cone which is about 1-2 centimeters diameter and a maximum temperature

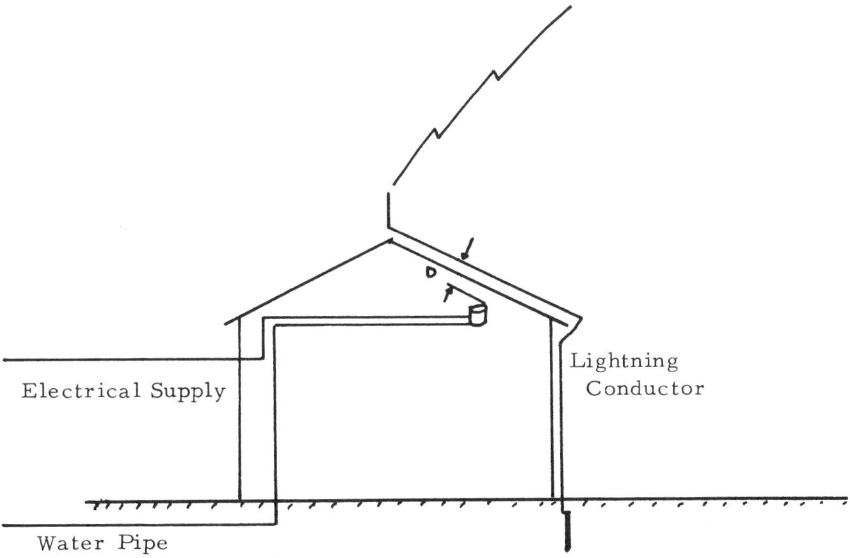

Figure 12. Lightning strike to house and conductor showing distance to interior grounded unit

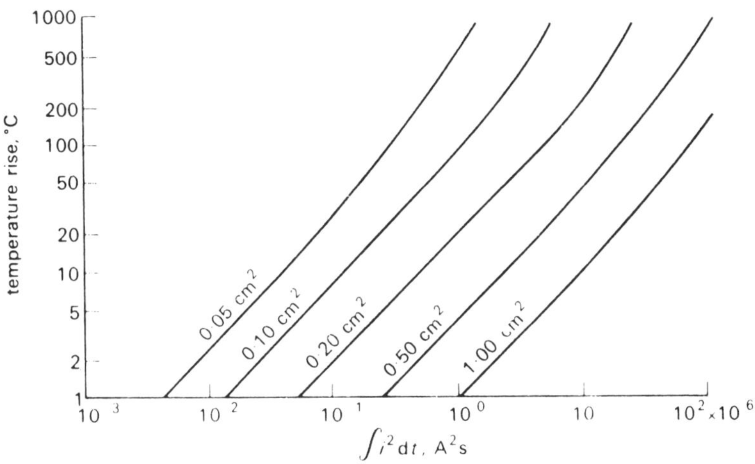

Figure 13. Temperature rise in copper conductors (after Golde (3))

of about 60,000 °F is reached after a few microseconds.

The effects of this temperature on metals is usually not serious, although there is a possibility that a thin metal sheet will be penetrated. The temperature rise is proportional to i^2t, where a maximum value of $\int i^2t\,dt$ is about 10^7 amp^2 sec. Neglecting dissipation and referring to Figure 13, one sees that the temperature rise of copper conductors, as specified in most lightning codes as 30 to 50 mm^2, is moderate. For aluminum the temperature values can be taken as 1.5 times those for copper.

There is one aspect of heat dissipation that needs consideration. Where the lightning current is being discharged through a high resistance joint, such as a poor contact or overlapping metal sheets, the heat generated may give rise to heavy sparking. Penetration may occur in the case of thin metal sheets such as used in roofing material or aircraft skin. The size of the hole is a function of lightning charge, the material and its thickness. For 20 mil copper, the hole could be up to 300 mm^2.

When lightning strikes an insulating material the point of contact could be raised to a high temperature and penetration could result. By these means clean holes of 2 cm diameter have been punched in glass by lightning discharges. If this insulant contains moisture the current will flow preferentially along the path of best conductivity. Moisture can be converted into steam and explosions occur. Enormous blocks of concrete have been demolished this way and on one occasion rocky ground was furrowed for 800 feet and 75 tons of rock and soil dislodged. The explosive effect was equivalent to 600 lbs of TNT.

Mechanical Considerations. Mechanical effects concern the shock wave and bending forces. With the rapid temperature increases discussed earlier, the air surrounding the channel expands extremely rapidly and produces a supersonic pressure wave. Figure 14 shows how this wave is propagated from the central cone, as calculated by Hill (7). It is responsible not only for thunder, but also for widespread lifting of tiles on the roofs of buildings.

Two parallel conductors caught in a lightning discharge are subject to attractive forces and these forces are responsible for the fusing of stranded conductors and for squashing hollow conductors.

There is one more mechanical force worth considering. If a lightning conductor follows a right-angle bend on a building and this conductor has to discharge a lightning current, it will be

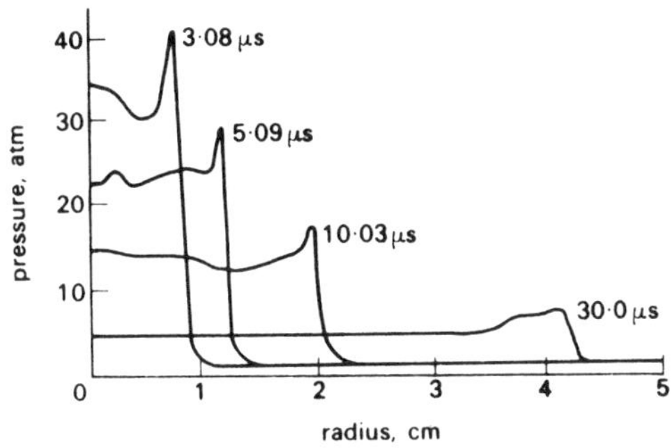

Figure 14. Development of pressure from lightning channel

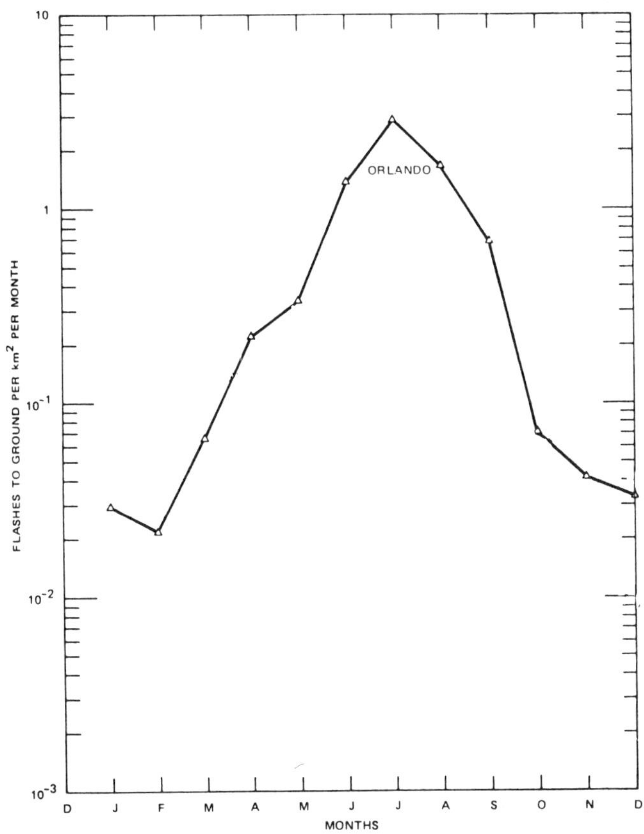

Figure 15. Annual variation of thunderstorm activity in terms of flashes to ground in Orlando, Florida

subject to a mechanical force trying to straighten it and thus, attempting to bend it outward. The magnitude of the force is proportional to the square of the current, but even for large strokes it can only reach about 5,000 lbs. Sharp rectangular bends in conductors should, therefore, be avoided.

Frequency of Strikes to Ground. The duration of a thunderstorm cell is of the order of 30 minutes, with a flashing rate of 2 or 3 per minute. Since a storm contains one or two active cells at any instant, 3 to 4 flashes per minute is not unreasonable as an average with a storm duration of approximately one hour. The proportion of discharges that go to ground is quite variable from storm to storm and also during phases of the same storm. Typically within the United States the ratio of cloud-to-ground over intracloud discharges is 1 to 4. It follows, therefore, that one per minute is an appropriate round figure estimate of the occurrence rate for flashes to ground with a possible occasional increase to one every three or four seconds in a severe case. Figure 15 illustrates the annual variation of discharges to ground for Orlando, as calculated by Cianos and Pierce (8).

The spatial distribution of flashes can be a significant factor in lightning problems. The fact that consecutive flashes are well separated, often quasi-randomly, is familiar to any careful observer of thunderstorms. Nevertheless, the erroneous concept that the discharges progress in a steady orderly pattern is still prevalent.

The only thunderstorm statistics which are readily available is the thunderstorm day. A day is defined meteorologically as a thunderstorm day if thunder is heard; this implies the oc currence of lightning within about 15 km of the observing site. No account is taken in the thunderstorm day statistic of the number of times thunder is heard, nor the number of discrete thunderstorm events per thunderstorm day. Figure 16 illustrates the number of thunderstorm days recorded in the USA, and showing an excess of 100 such days in the St. Petersburg to Orlando region of Florida with a value of about 80 at Cape Canaveral.

Cianos and Pierce give a useful relationship for determining the frequency of strikes under a thunderstorm. Two methods are available by which data for T_m can be converted into estimates of σ_m, the flash incidence per km^2 per month, and hence, the ground flash incidence ($p\sigma_m$) per km^2 per month, by multiplying by the proportion p of flashes that go to earth. The two methods

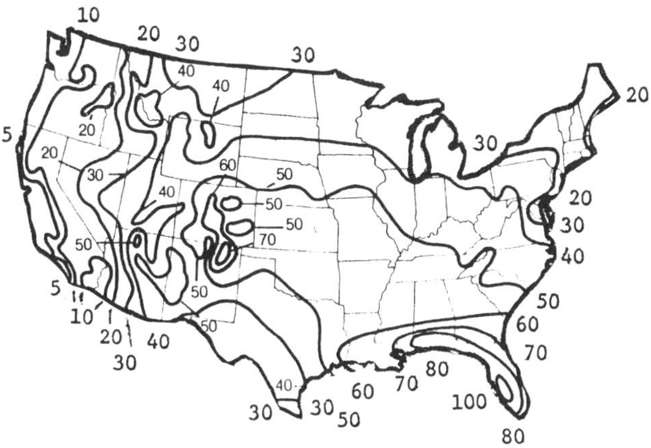

Figure 16. The average number of days per year on which thunder is heard in various parts of the USA

Table I. Monthly Dependence of Thunderstorm Activity

Cape Kennedy 1957-1962

Month		(p = 0.18)	
	T_m	σ_m	$p\ \sigma_m$
January	.5	.12	.02
February	1.8	.25	.05
March	3.7	.53	.10
April	3.3	.45	.08
May	6.7	1.42	.26
June	14.0	5.91	1.06
July	13.8	5.75	1.04
August	15.8	7.52	1.35
September	10.5	3.35	.60
October	3.8	.55	.10
November	0.7	.15	.03
December	0.7	.15	.03

do not differ greatly in the results they yield near the range (2 to 10) of T_m of most practical importance. Pierce concludes that:

$$\sigma_m^2 = a\,T_m + a^2\,T_m^4$$

where a equals 3×10^{-2}. Table 1 illustrates these results at Cape Kennedy where the values of p are taken as 0.18.

Using these formulas we conclude that approximately five discharges come to ground each year per square km.

Frequency of Strikes to Tall Structures. The incidence of strikes to tall structures electrically connected to ground is controlled by two factors. One is the attractive radius and the other is the triggering factor. Pierce and Price (9) have investigated these factors in detail and a few of their findings will be discussed here. The attractive radius, r_a, and its associated attractive area A_a ($= \pi r_a^2$) are primarily functions of structure height h. The attractive radius is defined as the average radius at which a downward leader from the cloud is just able to induce an upward streamer from the structure that will unite with the downward leader and thus divert the flash to the structure. The triggering factor represents the inclination of flashes to be initiated at the tip of the structure; it is negligible for $h \leqq 100$ m, but as h increases, triggered flashes become increasingly common and for $h \geqq 250$ m the triggered variety of discharge is by far the more important.

It is possible to calculate r_a. However, the calculations that have been made can be criticized in many respects. Cianos and Pierce (8) have given a complicated expression for r_a as a function of h. This is based both on the mathematical representations emerging from theoretical analysis, and on an empirical fit weighted according to the degree of reliability of the various data sources. Table 2 shows r_a as a function of h. Note that above about 150 m, the attractive radius does not change with a further height increase. This is because calculations indicate that for $h \geqq 150$ m, the field distribution between the tip of the structure and the downcoming leader is not much influenced by the presence of the ground.

Pierce has pointed out that reported instances of triggered lightning occur when the ambient general electric field E_a lies between 3 and 30 kV/m and the voltage discontinuity V_D, between the tip of the conductor causing the triggering and the unperturbed atmosphere, is 0.3 to 6 MV. It seems plausible that for the lower values of E_a or V_D there is a small but finite chance of lightning being triggered; this chance will obviously be greater the longer the values of E_a and V_D are maintained. As E_a and V_D increase so will the probability of triggered lightning, but the chance will again be dependent on the length of time for which any specific values of E_a and V_D exist.

Table 3 summarizes in Column 1 the best presently available data on the incidence of triggered lightning as a function of height. The data base is so scanty that substantial future modifications could occur. Also shown in Table 3 are the information derived from two expressions by Pierce and some theoretical results due to Horvath (10). None of the theoretical expressions agree well with the experimental data. Horvath's work much overestimates the incidence at lower values of h, and gives underestimates for high h. Expression (1) fits well for $h \leq 150$ m, but overestimates for large h.
Expression (2) underestimates throughout, but the agreement is becoming better for $h \sim 400$ m.

As an example, let us consider a 260 foot or 80 m tower at Cape Canaveral. Table 2 gives the attractive radius as 310 m and Table 3 indicates an average value of .16 for the ratio of triggered to natural lightning. The incidence of flashes to ground at the Cape has been shown in Table 1 to be $4.7/km^2$. Thus, the annual incidence of natural lightning to the tower should be,

$$4.7 \times \pi \times (310)^2 \times 10^{-6} = 1.42$$

Triggered lightning should contribute a further incidence of some,

$$0.16 \times 1.42 = .23$$

The total number of strikes to the tower will, therefore, be on the order of 1.65 per year.

Lightning and Switching Surges and Transients

Surge or transient voltages are one of the least understood elements of electrical energy for the simple reason that very little data is available on the subject. These unwanted sub-microsecond time-to-peak voltage transients of possibly several thousand volts and several hundred amps can exist in all systems. The random characteristics of their magnitude and duration make them difficult to identify and analyze. They appear unexpedtedly and their presence is still unsuspected, even after component damage or software default. The damage can be a rapid component failure or transients can slowly attack the component causing degradation, which can produce random

Table II. Relation Between Structure Height (h) and Attractive Radius (r_a)

h (m)	r_a (m)
25	~ 150
50	~ 250
100	~ 350
150	~ 400
> 150	~ 400

Table III. Proportion of Triggered to Natural Lightning

Structure Height (m)	Actual Data	Expression (1)	Expression (2)	Horvath Theory
50	~ 0	~ 0	~ 0	0.1
100	~ 0	~ 0	~ 0	0.2
150	0.3	~ 0	0.5	0.4
200	1	0.1	2.8	0.7
300	4	1.3	16	1.4
400	10	6	38	3.0

erroneous signals for several weeks until total component failure occurs.

The elimination of this rapid energetic surge requires special devices and can not be prevented by crowbars, gas diodes, or overvoltage protectors commonly found in computers and power supplies. Considerable problems exist with filtering devices, unless they are especially made with much higher than expected insulation requirements and are used in conjunction with fast, specifically designed overvoltage protectors.

Evidence exists that transients entering computer systems often cause software problems by causing bits to be changed and bits to be added to memory locations. These transients have also been shown to be caused by computer tape recorders providing the inductive impulse.

There are three specific reasons for the change in voltage suppression requirements: (1) The increased effect of induced lightning on solid-state components (also enhanced by a changing world weather pattern). (2) More and more sophistication in semiconductor technology, simply smaller and smaller devices. (3) The very monumental effect of switching transients/surges caused by a degenerating supply of commercial power. As the demand upon power companies increases, at a geometric rate, the ability to produce power does not increase at the same rate. Therefore, loads are constantly being switched from one line to another, causing "surges" (which in turn, cause high speed, short duration transients to proceed down the power line). As little as one (1) nanojoule (1×10^{-9}) of energy applied to the semiconductor can cause a shut down of operations. Figure 17 is a computation by Odenberg (11) showing how a small amount of energy can either upset or destroy a transistor, I. C. or semiconductor.

The major sources of lightning surges in conductors are due to

a) Ground potentials caused by nearby lightning strokes;
b) Induced effects caused by lightning current flowing on a shield;
c) Direct strikes to a wire;
d) Side-flashes to the conductor from a nearby strike;
e) A straight conductor acting as an electrical field change antenna for lightning effects;
f) A looped conductor acting as a magnetic field antenna for lightning effects.

Burying the cable does not remove lightning effects, as the cable is then an ideal ground path for the current. The lightning current may side-flash several meters to the conductor under the ground, where the distance is primarily a function of solid resistivity and the resistance of the conductor to ground.

The largest lightning voltage recorded on a transmission line reached a peak value of 5 million volts in less than two microseconds. The resulting oscilloscope recording is shown in Figure18, and the strike occurred some 4 miles up the line. It is suggested that closer to the strike point the current rate of rise was probably of the order of 10 million volts per microsecond.

Residential 120V AC lines are found to experience peak lightning associated voltages of up to 6kV and internal switching transients up to 3kV. The transients will be oscillatory in nature, with a fundamental frequency from a few tens of kilohertz to several megahertz with components ranging into hundreds of megahertz. They will last from 100 nanoseconds to 100 microseconds and can be clamped within a few cycles. Good grounding and bonding may reduce the transients significantly.

Intracloud lightning causes a considerable number of induced effects in cables of several thousand volts and several hundred amps, even though the separation distance of cable to discharge may be several miles. The main reason for such an effect is that the power, telephone or data cable acts as an antenna. Shorter cables give rise to larger surges due to reflections at the cable ends.

Transients from switching inductive leads, i.e., solenoids and relays, and local motors, such as air conditioners and heaters, can cause several thousand surges per day. At an Air Force site over 53,000 damaging surges over 1 joule were recorded during a one month period, of which 10,000 occurred in one day (Odenburg).

An example of induced voltages would be a peak of "3900 volts" produced by a 4.0 amp, 30 volt inductive load (Aeronautical Radio, 1967). The Navy regard a 2.5 thousand volt peak voltage as a maximum inductive switching transient on 110 VAC systems. This would be generated by large inductive switching, like elevators in hospitals. The American National Standards Institute says that 110 volt line faults can cause six thousand volt transients. Sparks from a charged human can also reach two thousand volts in one nanosecond and several tens of thousands of volts shortly after.

The utilitie's induce surges while energizing and de-energizing transformers in an effort to manage loads. Although utilitie's provide primary protection, the front end of surges may

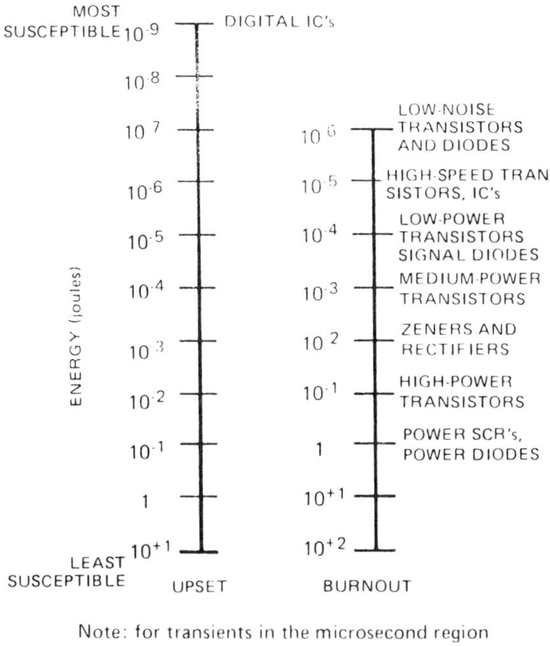

Figure 18. Oscillogram of voltage surge on a 110kV transmission line

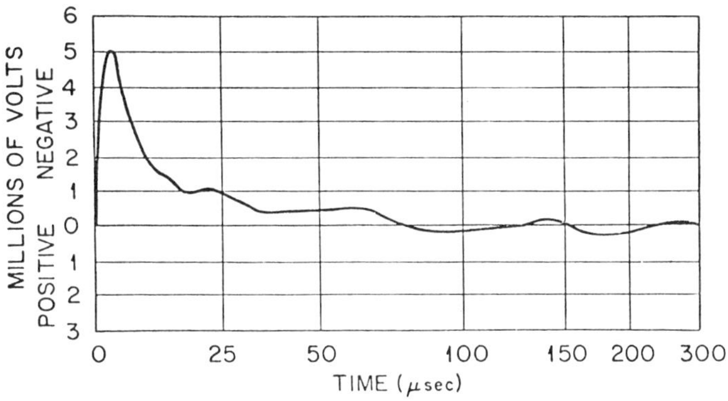

Figure 17. Upset and burnout energies for various semiconductors

not be eliminated. Also, the energy stored in transformers de-energized after the power company's arrestors fire and remove power for 1/2 cycle can also cause damaging surges. A surge also appears at power re-initialization.

Another important source of surges is human error. While attempting to wire, adjust or service systems, unwanted transients are produced which may prove to be detrimental to semiconductors.

Insulation Effects. The effect of high voltages on insulation can be quite large and whereas the initial breakdown of an insulator may not be catastrophic, the repetitive effects of high voltage transients will produce breakdown at the same place until the insulator cannot even stand the steady state voltage. An electric clock manufacturer reduced his failure rate to one hundredth of his earlier failure rate by increasing the insulation level from 2 to 6kV. Surge protection would have the same effect.

Breakdown will also occur along a surface such as a printed circuit board. In this case, a path of slightly conductive carbonized insulation will occur which may also be influenced by vaporized metal. Steep wavefront voltages may lead to breakdown of insulation between the windings of a coil.

Grounding and Bonding. When one grounded conductor is conducting a steep wavefront current, large potentials may develop between it and another grounded object. The effects of this problem have been illustrated earlier. Bonding the grounds together would have removed the problem. The inductive effects of such bonding between two objects must be carried out with care or it may still lead to hazardous potential differences. This problem is illustrated in Figure 19, where a cable shield and an enclosure are connected together (12). This example indicates that ground leads should be kept as short and direct as possible. Figure 19a shows a shield being terminated on an enclosure which allows the current to flow radially from the shield to the enclosure. If a separate ground lead is used to carry surge current, it will have an inductance (Figure 19b) which will lead to a potential difference between shield and enclosure. The inductance of one inch of 0.034" diameter wire is about 0.0134 μh and with a rapid current pulse of 100 amp/nsec, the voltage developed will be 1340 volts. Ten inches of wire will enable 13k volts to develop between shield and enclosure, implying a spark over 1 cm long. The wire should, therefore, be kept as short as possible.

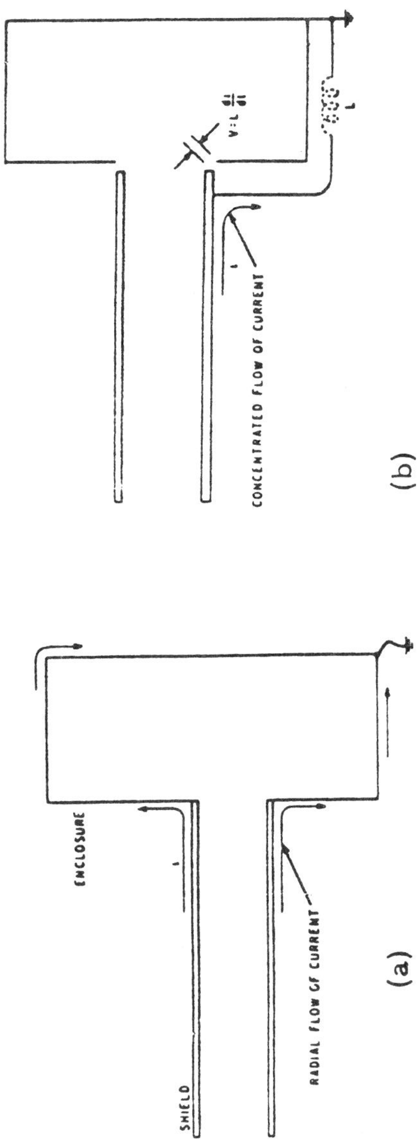

Figure 19. *Connections between shield and enclosure; (a) good, (b) bad*

It is also absolutely essential that a shield be grounded at both ends in short lines. The magnetic fields caused by lightning can induce voltages around open circuit loops and currents around short circuit loops. The induced currents hardly ever cause damage, but the induced voltages can be excessively high and will cause damage.

Ideally, one must set up separate grounding systems for the various electrical parts of a system and combine these separate grounds at only one common reference point.

Protector Devices

Protector devices are of two basic types, constant voltage and crowbar. The constant voltage devices will conduct very little at the steady state voltage, but above a certain voltage level will conduct very heavily. A crowbar device in effect, short circuits a high voltage to ground. This short will continue until the current is brought to a low level. A constant voltage unit will never reduce the line voltage below its steady state value, but the crowbar device often will. This could be a problem if there is a continuing follow current.

Constant voltage devices in everyday use are avalanche and zener diodes and varistors, or voltage dependent resistors. Spark gaps and gas discharge tubes are the most common type of crowbar.

Low pass filters are often used as suppression devices. A capacitor placed across the terminals is the simplest form of filter, where the impedence it should present to the transient will be much lower than the transient source impedence. This approach will work well unless the capacitor loads down the desired voltage and does not create current in-rush problems. A resistor in series will help, but will reduce the effectiveness of the filter. A capacitor network is also ineffective if the transient has high energy in either polarity. Filters can become expensive and must be very carefully designed.

Isolation transformers may allow surges to be capacitively coupled across the windings, thereby passing them into the system. At times the load may be such that the input surge is differentiated at the isolation transformer, causing a much more rapid risetime and hence, making it more dangerous. Considerable thought and calculation must be performed before deciding to protect by such a transformer.

The lead length of suppression devices can cause large overshoot voltages depending on the rate of rise of the current.

It was discussed earlier that a one inch wire can lead to a voltage overshoot of over 1300 volts. It is possible, however, to purchase most protection devices in disc form without wires, and with careful mounting the voltage overshoot will be negligible.

For selection of protection components and devices, the transient voltage withstand level of equipments and circuit components must be known or conservatively estimated. For most equipment and circuit components, standards do not exist for lightning transient levels. Therefore, information available from manufacturers must be obtained, laboratory testing performed or conservative engineering estimates used. Limits for common types of equipments and components are provided for guidance.

- Transistors and Integrated Circuits: 2 times normal voltage.
- Diodes: 1.5 times their peak inverse voltage.
- Small motors, small transformers and light machinery: 10 times normal operating voltage.
- Large motors, large transformers and heavy machinery: 20 times normal operating voltage.

In addition to the above, capacitors are many times overlooked and unless their dielectric punch-through voltage for transients is known, limiting transients to 1.5 times the DC working voltage is recommended.

As an indication of line transient/surges, a commercial power line of 480 volts to a satellite tracking station was monitored for spikes greater than one joule by Odenburg (11). The number of surges recorded is displayed in Figure 20. It is interesting to note that the 53,020 surges monitored in September alone is in keeping with other published data. At times 10,000 transients have been recorded in a 24 hour period, many of them occurring within millisecond periods.

Avalanche Diodes and Zeners. The volt-ampere characteristics of a semiconductor diode are shown in Figure 21 in which there are three principle regions of operation. The forward biased region is limited by the external circuit, and the leakage region is where the voltage is reversed, but it is still less than the critical value. When the reverse voltage increases beyond this critical value, the reverse current increases sharply and the diode is operating in the breakdown region. Normal rectifier diodes are made to operate in the forward and reverse-biased region, but transient suppressor operate around the breakdown region.

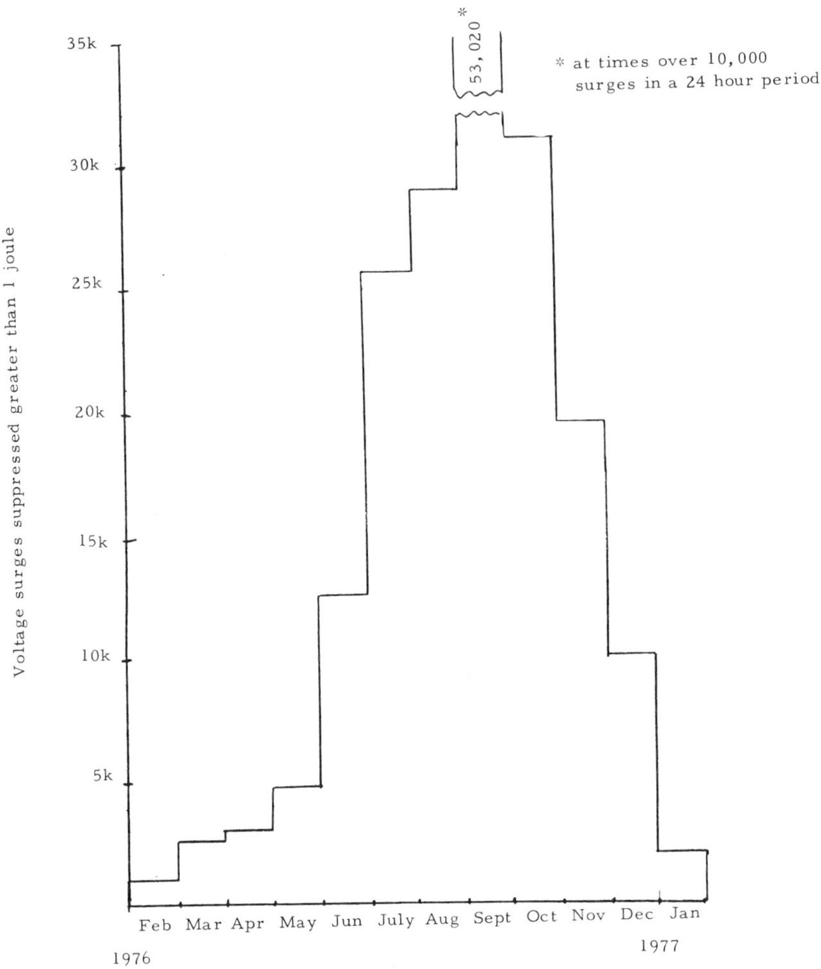

Figure 20. Surges greater than one joule suppressed at a USAF site, Odenburg

Avalanche diodes exhibit a sharp turn at the knee, but zener diodes go through this transition more gradually. This implies that the avalanche diode is a better suppressor for transients than zener diodes.

These devices are the most "constant voltage" devices available and the voltage is only slightly dependent on the current. The operation takes place within a very small volume of silicon, where the energy or heat generated by the transient can cause failure at the junction. Special suppression devices using the silicon avalanche technique are manufactured that have a junction area over ten times larger than a one watt zener diode. These devices will clamp at speeds in excess of 10^{-12} sec and, depending on the size, the peak power rating can be up to several hundred kW for a 1 μsec pulse. They do have a small capacitance, however, but with careful design it is possible to use them in protection circuits at frequencies in excess of 100 MHz.

Varistors-Voltage Dependent Resistors. A varistor is a bulk semiconductor device whose resistance varies with the magnitude but not the polarity of the applied voltage. Varistors are composed of a polycrystalline material made by pressing and heating special mixtures containing either silicon carbide (SiC) or oxides of zinc and bismuth. Metal-oxide varistors (MOV's) have a more nonlinear V-I relationship and therefore better clamping. They are highly nonlinear elements developed recently for protection of electric devices from induced voltage surges. In the absence of abnormal voltages, the MOV presents a very high resistance at its terminals; however, in the presence of a surge, its resistance diminishes by several orders of magnitude, thus absorbing the energy of the transient above a specified value.

MOV's provide low voltage nonlinear elements with voltage-current characteristics worse than zener diodes, but with a bi-polar property and high energy dissipation/size capability. These devices, primarily intended for surge protection of AC power lines, will be also applicable to low voltage signal line protection when lower voltage types of MOV become available. The step response of an MOV is in the 50 nanosecond region. Typical V-I curves are shown in Figure 22.

Gas Breakdown Devices. On the opposite end of the ruggedness spectrum are the spark gaps and gas-discharge tubes (GDT's). These depend on the formation of an ionoized gas

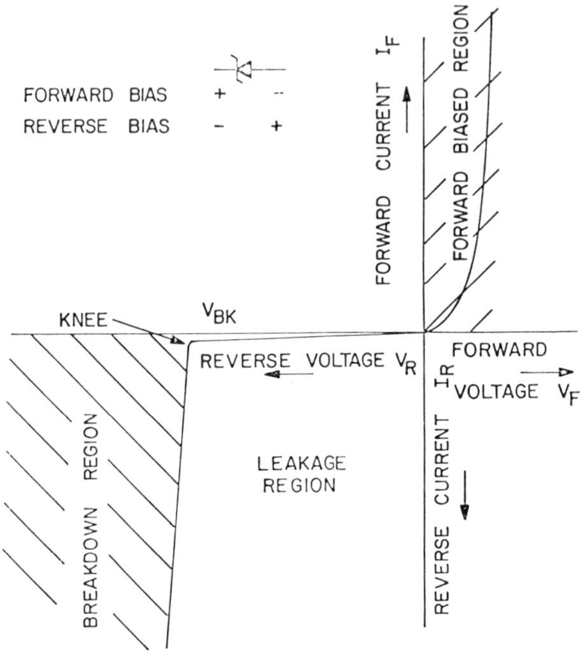

Figure 21. Schematic volt–ampere characteristic curve for a semiconductor diode

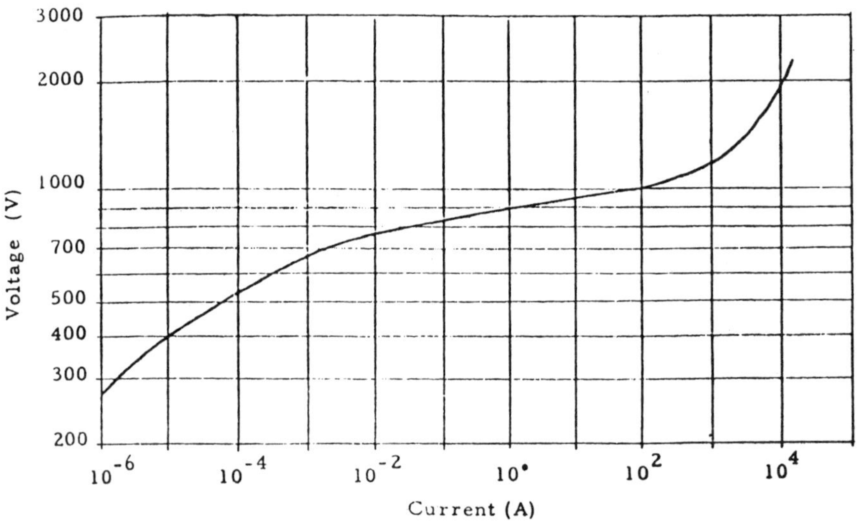

Figure 22. Volt–intensity curve for MOV

between metal electrodes. The gap length, gas pressure, and several other factors determine the breakdown voltage. When an arc is formed, the suppressor is capable of conducting high currents at a low voltage ($\lesssim$ 100V).

Unfortunately, the steady state power source is frequently capable of keeping the arc conducting until current and voltage are reduced, temporarily disabling the supply. Many such suppressors also have a noticeable response time, such that a fast-rising transient reaches a high voltage before the arc can form. GDT's are not generally feasible below 90 volts.

Druyvesteyn and Penning (13) describe the action of a gas discharge device in Figure 23, where one can see the important glow and arc discharge regions. The arc discharge passing high currents at low voltage. An excellent description of the use of spark gaps is given in a report by Hart and Higgins (14), and their main conclusions will be briefly described here.

Typical volt-time curves of a GDT are shown in Figure 24 indicating an initial high clamping voltage. Its use in the protection of an AC line surge is shown in Figure 25, but because the arc region can be sustained at a low voltage, the AC voltage may be sufficient to allow a follow or holdover current. This holdover current, depending on the power source, may be significant and may be sufficient to cause damage to the electrodes. As the voltage passes through zero at the end of every half cycle, the GDT will extinguish but at times, if the electrodes are hot and the gas ionized, it may re-ignite on the next half cycle. The gas tube is an excellent device for protecting against high current surges, but can not be used effectively in protecting low input impedance circuits. It is often an advantage to provide added protection to clamp the initial voltage overshoot that the GDT is not capable of protecting against. This can be done in several ways by designing hybrid circuits with the gas tube as the initial current protector and a solid-state device to protect against the initial overshoot.

Results from Protected Systems

During the last two years Atlantic Scientific Corporation have been working with a large Ohio based company that manufactures brain and body scanners for hospitals. These are commonly called CAT scanners (Computerized Axial Tomography), and each consists of a large computer and several other microprocessor systems.

It was common for some systems to have occasional badly

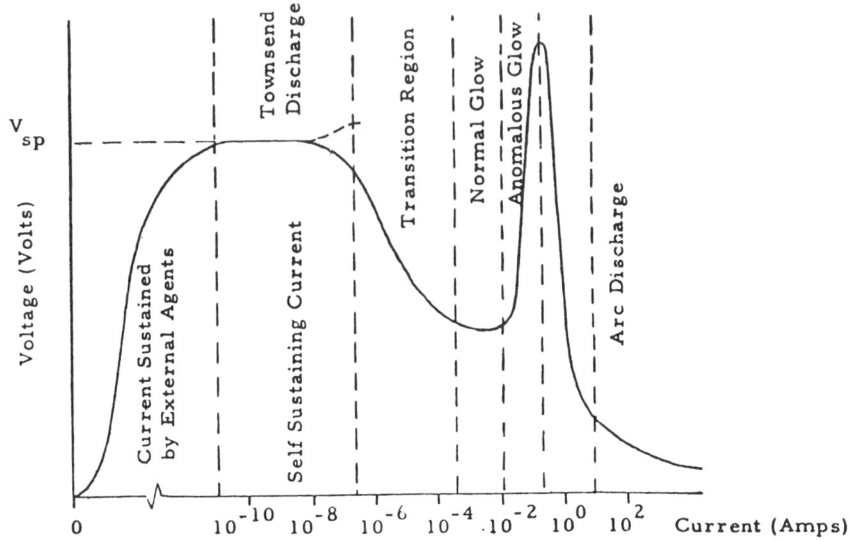

Figure 23. Schematic volt–intensity characteristics of a gas discharge between flat, parallel electrodes

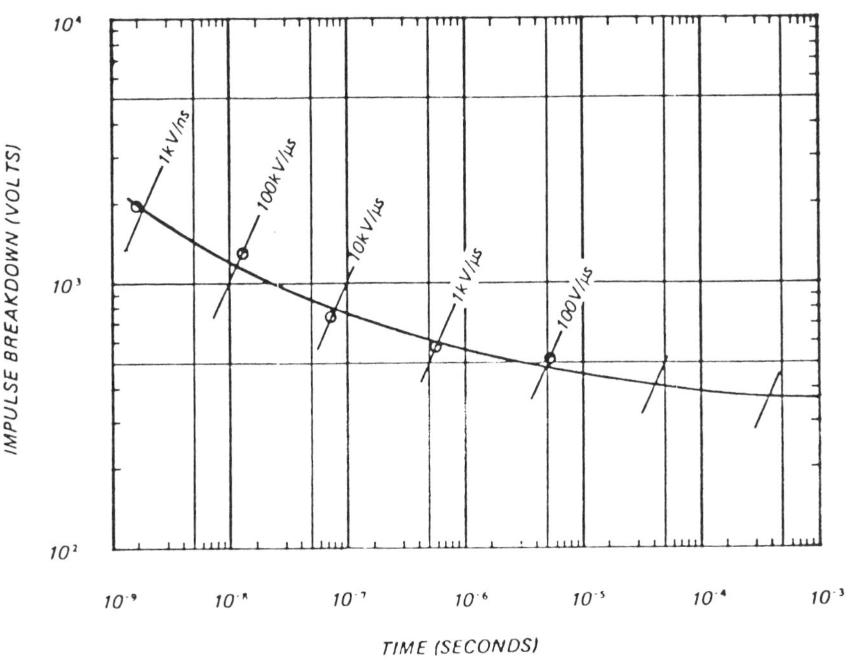

Figure 24. Small spark gap characteristics, Joslyn, 2301-14.

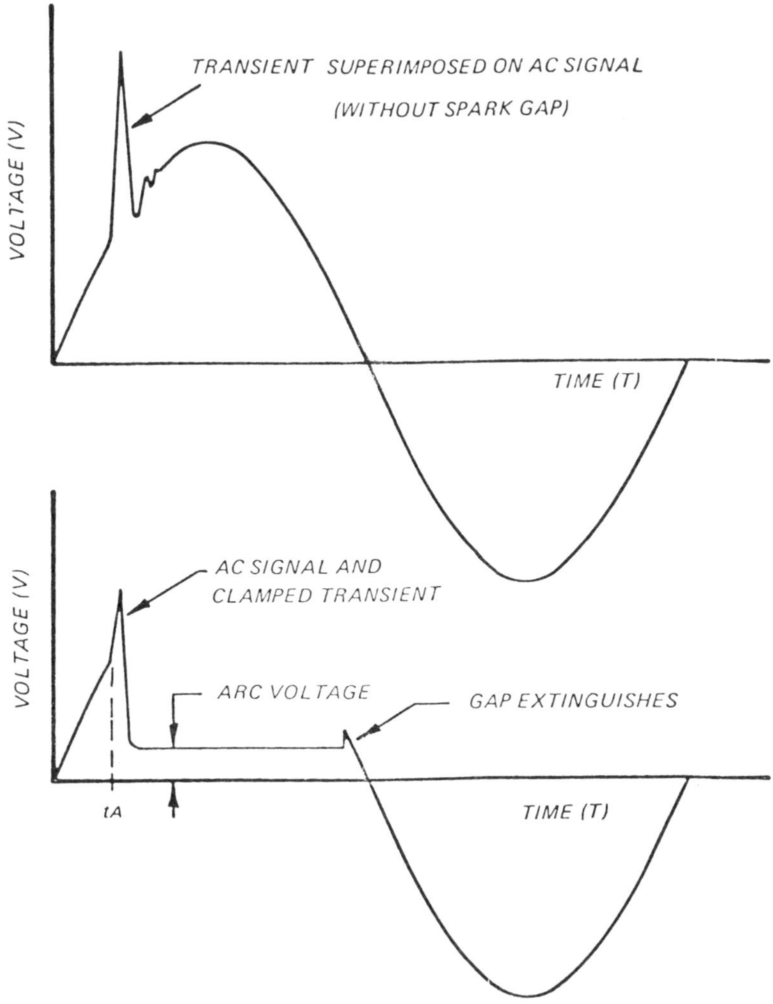

Figure 25. Volt–time curves of transient with and without spark gap protection

formed images or for the computer to suddenly issue false instructions during the course of a scan. Several units were provided with solid-state surge and transient suppression. Those ranged from front end devices capable of controlling surges of 25,000 A, to low voltage circuit protectors capable of clamping in picosecond times (10^{-12} sec).

Surprising results were obtained in terms of system downtime, system damage costs and scan quality. One hospital in Denver, Colorado indicated that many of their software problems dissappeared. These were false indications that a scan had been completed and also occasional automatic patient table movement when not asked for. A more remarkable feature was the improved performance in system downtime and electrical cost savings. The graphs shown in Figure 26 show the percent downtime for two Florida systems, with virtually zero downtime since kit installation. The P.C. board repair cost savings on one of these sites was over $4,000 per month. Since that time, over 190 systems have been retrofit with solid-state surge protection. Many of the units were designed for ease of installation and needed only separation of a plug and socket to allow a prebuilt connector/protector combination to be incorporated.

Unfortunately, the protector provided in many OEM computers and power supplies is for overvoltage or overcurrent protection for failing components, and its response time is far too long for the rapidly rising surges and transients entering the system.

The results from this study on the CAT scanners indicates that systems can be adequately protected, but caution must be placed on the performance of any off-the-shelf black box. A knowledge of its contents is a must in order to allow correct device selection and provide the adequate protection required.

Lightning Warning and Tracking Instrumentation

Most instruments and techniques for lightning warning, function by monitoring the range or location at which lightning is occurring and an excellent comprehensive description of the majority of them is given by Cianos and Pierce (8).

One of the more justifiably acceptable approaches until recently has been a combination of field and field change recordings. The first design of a warning device combining these two techniques was built in 1962; this employed a corona point to measure the field, and a circuit to detect electric field changes. The combination approach is necessary as there are two aspects of lightning warning that are important. One is the detection of existing lightning, and the other is the detection of a charged cloud that is about to discharge itself to ground for the first time.

The most modern and satisfactory method of monitoring the charge build up in a cloud is to use a field mill to measure the electric field below it. When lightning develops in the cloud, there is an electrostatic field change that can be monitored to give an estimate of flash distance. Any corona current indication is affected by wind speed.

Field Mill. Fair weather electric fields are typically measured at a low positive value of + 100 V/m, and storm conditions commonly produce high fields in the range of ±5,000 V/m. When a storm is moving in, the very sudden excursions in the field due to cloud discharging can be easily recognized. When the charge is building up in a cloud overhead, a change in the polarity of the electric field can be observed from the positive fair weather value to a high negative value. For such situations, a field of -2,000 V/m can be used as a first order estimate that lightning is possible. However, such a fixed value should not be relied on solely for determining warning levels; studying the development and build up of the electric field over a period of time can give much more accurate information. There are problems with peculiarities of storm fields and concealing of true fields that make a fixed value warning unreliable.

Considering overhead and more distant clouds, the field mill reading responds mostly to the fields due to the strong negative charge center in the lower part of the cloud. Depending on relative distances and angular relationships though, the small positive charge center at the very base of the cloud could sometimes have a dominating effect, and so could the large positive charge in the upper part of a more distant cloud. Such variable

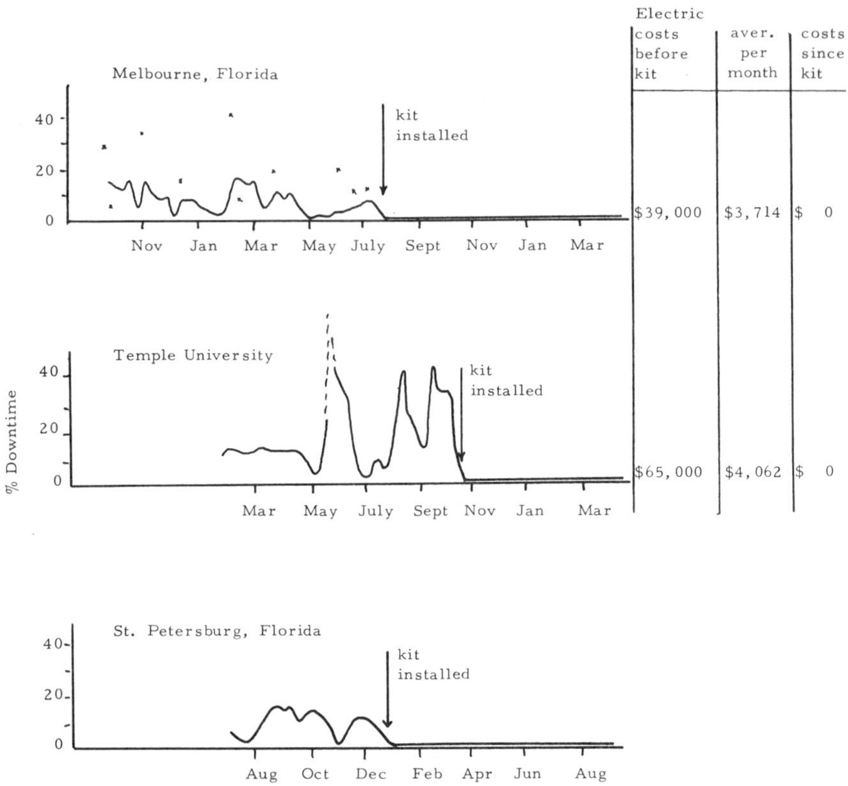

Figure 26. Advantages of surge protection on large system

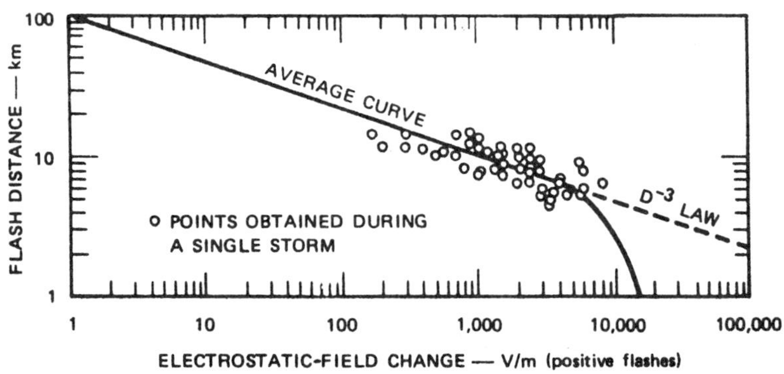

Figure 27. Electrostatic field change from lightning as a function of flash distance

conditions have to be recognized in order to achieve accurate interpretations of the field mill data.

The major problem in determining correct field values is caused by space charge screening. Ion clouds can be formed by corona given off from sharp points on structures and natural vegetation sources under high fields, also by ions in exhaust fumes, etc. Such charged regions close to a field mill can mask the effect of the cloud overhead, and produce false field indications. To obtain a more reliable determination of charge build up, a field mill should be placed above much of the corona on top of a large round structure such as a water tower. The field at such a location would only be enhanced by about a factor of 3.

Field Change Equipment. The instrument counts lightning flashes and operates essentially by relating the size of the change in the electric field caused by lightning to the distance between the receiver antenna and the lightning flash. It also tends to discriminate in favor of ground discharges. This field change is on average, 3 V/m at 40 miles and 5,000 V/m at 3 miles. Flashes are not of identical intensity, and if some very large current strokes occur at a great distance, then a false indication of a somewhat closer stroke will be given. Considering however the selectiveness of the equipment and the physics of the problem, it is shown by the following arguments that the uncertainty in the distance determination is not so great.

The instrument should be designed to operate in the frequency range below 1 kHz, and thus be sensitive to lightning over a distance from 0 to 100 miles. This determines that the equipment responds essentially to the near field electrostatic component of the electric field change, which for these conditions is dominant over the radiation component. The radiation component of the electric field change varies with $(1/\text{distance})^3$ and hence, the distance determination is quite reliable. The $1/d^3$ relationship is based on the approximation that the length of the lightning channel is small compared with its distance from the observation point. Within a 10km range, this does not always hold true and hence, deviations from the $1/d^3$ relation can be expected. Figure 27 illustrates some electrostatic field change measurements which follow the theoretical curve closely.

Using this type of equipment as the basic sensor, Atlantic Scientific Corporation developed a new instrument that integrates the number of lightning counts over a variable time interval from 10 to 60 seconds and presents a digital display of the number of counts separated in four distance ranges, 0-5 miles, 5-10 miles,

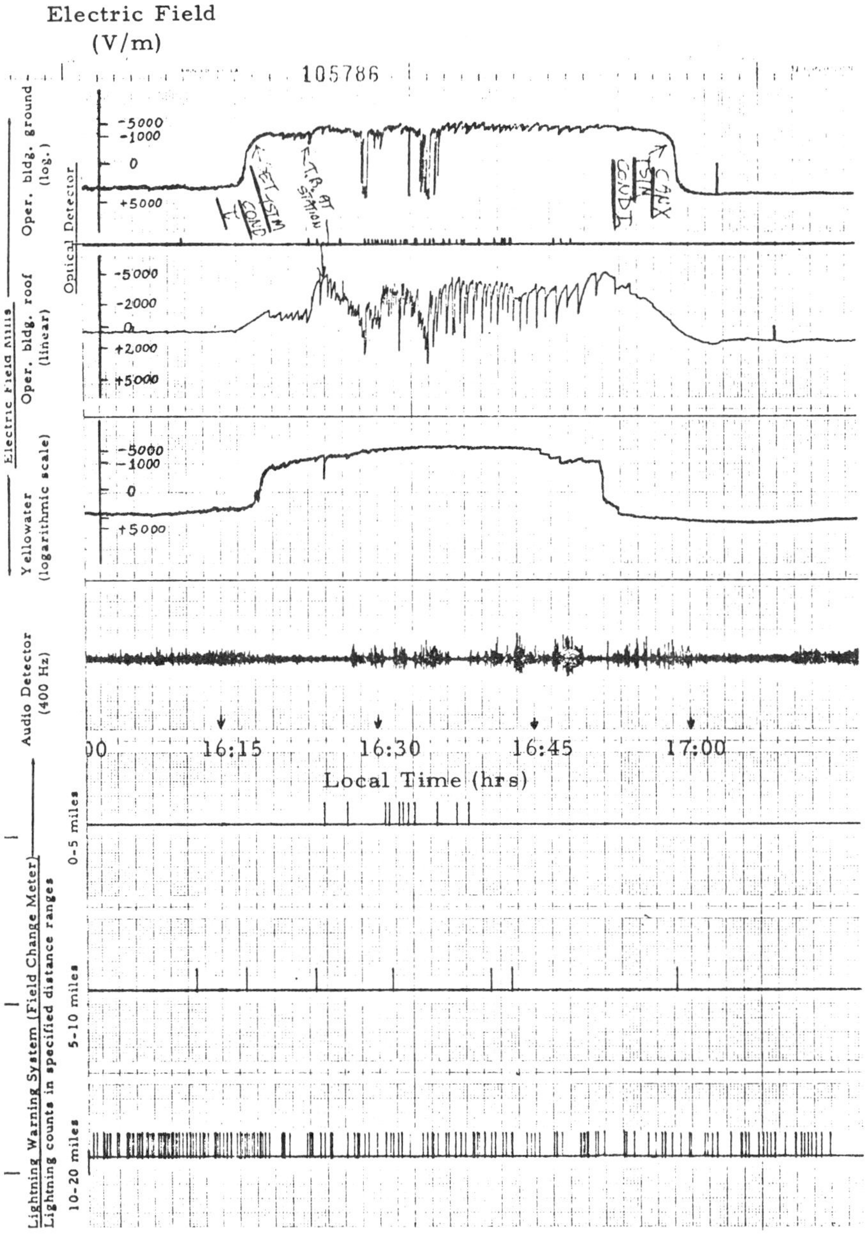

Figure 28. Results for lightning warning instrumentation at a naval station. The top three graphs are of electric field and the lower three of electric field change excursions greater than three preset values.

10-20 miles and 20-100 miles. An integration time of one minute makes erroneous distance readings easier to recognize, since the integrated count at the incorrect distance is much smaller than the simultaneous count in the correct distance range. An audible warning that can be set at any range level is also included. In addition, an output of each lightning strike separated into the four distance ranges was provided for the recorder. Figure 28 illustrates some results of this equipment showing electric field charge build up and warning several minutes prior to close lightning.

Lightning Position and Tracking. An extremely sophisticated Lightning Position and Tracking System (LPATS) has been designed to locate and track thunderstorms out to 300 or 400 miles with an accuracy of better than 3 degrees.

Lightning location techniques by triangulation beyond 150 km have been in existance around the globe since World War II which provide good accuracy, but for close lightning the tortuous channel as well as ionospheric reflections lead to large errors. LPATS relies on being able to detect the cloud to ground discharge by its unique broadband magnetic field waveform. Once detected, this waveform is sampled for the part of the return stroke that is within 100 feet of the ground. It is well known that this part of the discharge is almost always vertical and carries the greatest energy, implying that we have a vertical omni-directional radiating antenna and a powerful transmitter.

The system monitors ground stroke location, storm center and speed and direction of movement, as well as storm intensity and its variations. It can resolve multiple storms and is capable of displaying the information in map form on a TV screen. Additional advantages of the system are its capability of monitoring "hot" lightning which is known to start forest fires, and its ability to investigate its own performance and malfunctions.

ABSTRACT

A general understanding of the basic lightning process can lead to a much better understanding of lightning protection techniques and the resulting level of protection. The design of satisfactory lightning protection systems can only be achieved with a thorough knowledge of the mechanism and characteristics of a lightning strike, and the related problems that a steep voltage wavefront has on inadequate bonding and grounding.

Lightning induced line surges can also cause major damage to electrical or electronic systems such as computers, and it may also change data or programs without any permanent damage. The resulting effects can be disastrous where chemical mixing is governed by electronic techniques. A considerable portion of the damage caused by such transients and surges can be eliminated with careful planning of protection equipment, but the advent of solid-state components has placed considerable emphasis on the term "careful planning".

This paper discusses all these points and also attempts to educate the reader in lightning protection and statistics as well as lightning warning systems.

References

1. Llewellyn, S. K., Broadband magnetic waveforms radiated from lightning, M. S. Thesis, Florida Institute of Technology, Melbourne, Florida, 1977.

2. Anderson, J. G. and K. O. Tangen, Insulation of switching-surge voltages, in EHV Transmission Line Reference Book, Edison Electric Inst., New York, 1968.

3. Golde, R. H., Lightning protection, Edward Arnold Ltd, London, 1973.

4. Office of Naval Research, Code 450, Review of lightning protection technology for tall structures, Conference proceedings, 1975.

5. Bent, R. B. and S. K. Llewellyn, An investigation of the lightning elimination and strike reduction properties of dissipation arrays, Report No. FAA-RD-77-19, 1976.

6. Smith, R. S., Lightning protection for facilities housing electronic equipment, FAA-RD-77-84, May, 1977.

7. Hill, R. D., Thunderbolts, Endeavour, 31, No. 112, 3-9, 1972.

8. Cianos, N. and E. T. Pierce, Methods for lightning warning and avoidance, SRI Tech. Report 1, 1974.

9. Pierce, E. T. and Price, G. H., Natural electrical effects on the operation of tethered balloon systems. Stanford Research Institute Project 3058, March, 1974.

10. Horvath, T., Gleichwertige Fläche und relative Einschlagsgefahr als charakteristische Ausdrücke des Schutzeffektes von Blitzableitern. Int. Blitzschutzkonferenz, Munich, 1971.

11. Odenberg, R., Protecting facilities from induced lightning and power line switching transients, FAA-RD-77-84, May, 1977.

12. Fisher, F. A., Instruction Bulletin 53E 9007, Fischer and Porter Co., Warminster, Penn., 1970.

13. Druyvesteyn, M. J. and F. M. Penning, Electrical discharges in gases , Rev. Mod. Physics, 12, p. 87, 1940.

14. Hart, W. C., and D. F. Higgins, A guide to the use of spark gaps for electromagnetic pulse (EMP) protection, Report No. JES-198-1M-11/75, Joslyn Electronic Systems, Goleta, Ca., 1972.

RECEIVED November 22, 1978.

5

A Modern Propellant and Propulsion Research and Development Facility

WALTER W. WHARTON

U.S. Army Missile Research and Development Command (MIRADCOM), Redstone Arsenal, AL 35809

Modern propulsion and propellant systems require high performance. High performance usually precludes the use of docile, safe chemicals. Strong oxidizers, strong reducing agents, toxic and flammable substances, and explosive compositions are standard fare. Ammonium perchlorate, nitroglycerin, RDX and HMX (explosive nitramines), and nitrocellulose are typical ingredients used as oxidizers that also have explosive characteristics. Trichloroethylene, benzene, red fuming nitric acid, unsymmetrical dimethyl hydrazine and selected boron hydrides are examples of toxic chemicals that must be used in the development and production of rocket motors. The key word for most ingredients is reactive. Thus, in addition to being explosive and toxic, the substances generally will have other characteristics such as being corrosive, flammable, possibly carcinogenic, and a general tendency to be incompatible with other ingredients and materials of construction. To produce effective propulsion systems, facilities and procedures must be developed to synthensize, characterize and process such ingredients safely. A description of such facilities and procedures is given in the following paragraphs.

The Work Flow Process

Figure 1 outlines the work flow process. Each block represents one of the major topics addressed in propulsion system development. In each block is listed key considerations or operations that must be considered.

Ingredient Selection. Safety enters into consideration before the new molecule is synthesized. A new substance whose structure is known to be extremely sensitive, highly carcinogenic or flammable to the extent of being pyrophoric would not be considered seriously if any other molecule will suffice. Potentially applicable new molecules are submitted to the Army Environmental Hygiene Agency for toxicity evaluation. Several

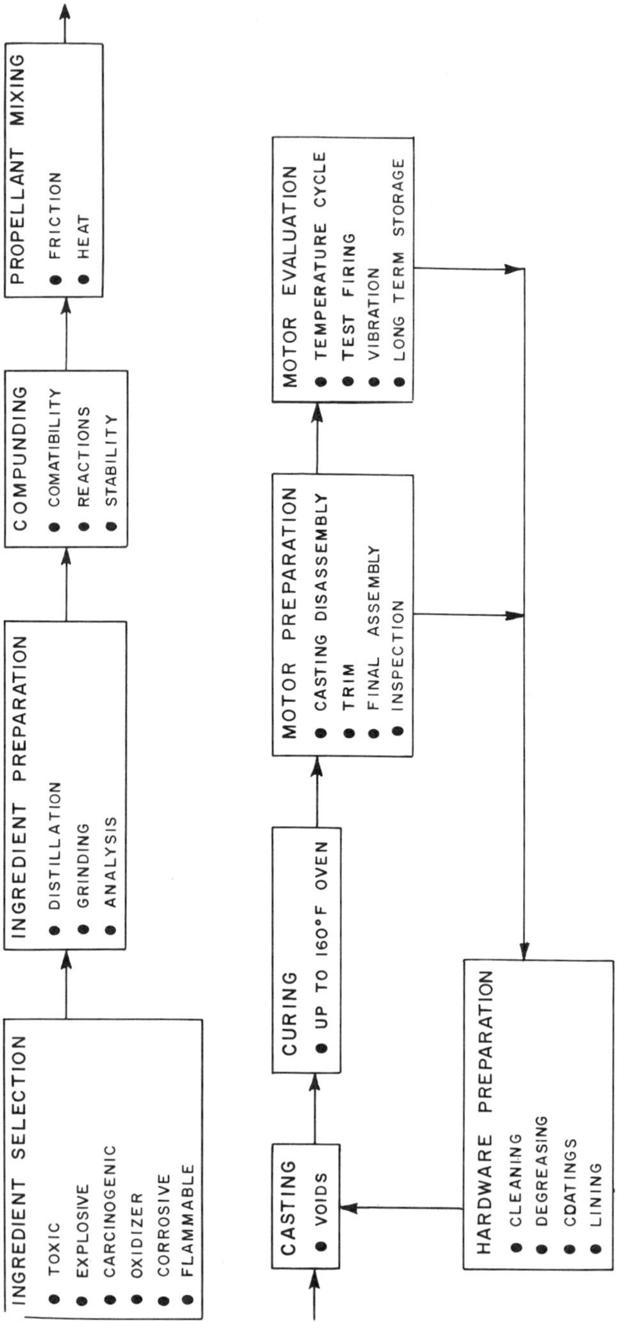

Figure 1. Work flow process

standard tests are utilized early in the evaluation phase to evaluate flammability, ignition and explosive characteristics. These include differential thermal analysis, thermo gravometric analysis, drop weight tests, friction tests, card gap (shock initiation) tests, and materials compatibility tests. Information derived from the above tests serve as a basis to establish safe procedures and techniques to handle and process the chemicals into propellants.

Ingredient Preparation. Sensitive ingredients such as nitroglycerin are frequently stored or shipped diluted with a solvent. Before use in an experimental composition, distillation of the solvent and preparation of the material is necessary. Solid oxidizers such as ammonium perchlorate (AP) require grinding to particular particle size distributions in order to achieve desired propellant combustion characteristics and required propellant mechanical properties. Such operations are performed in isolation. The SWECO building, figure 2, houses an AP grinder, Techniques for analysis of the ingredients and assay of the ingredients when compounded into propellants must be developed. Processes for sampling, handling and reacting that are safe and reliable must be devised.

Compounding. Modern solid propellants are highly filled binders (85-95% by wt.) and contain from five to ten chemical compounds blended together in a composition with a final characteristic somewhat like an automobile tire but with the added feature that it will sustain combustion in a closed system and under improper conditions may detonate. Propellants contain fuel molecules (usually a low mol. wt. liquid polymer), oxidizer molecules, burning rate modifiers, chemical stabilizers, combustion stabilizers, plasticizers, curing agents and binders (frequently the same as or part of the fuel molecules). They may contain other ingredients to impart specific characteristics such as enhanced tensile strength, reduced flash and contrail emission, or inhibitors to circumvent the effects of slow long term decomposition of unstable molecules. The objectives in this phase of the effort are to investigate and understand the interactions between the various components of the propellant, understand their compatibility with materials of construction, and understand their long term slow interactions which may lead to propellant deterioration and hazardous situations. Gross incompatibility (e.g. immediate obvious reactions, explosions, ignition) are immediately discovered and avoided. A continuous effort is maintained to investigate the chemistry propellants undergo during processing, curing, long term storage, exposure to temperature cycling, and to a limited extent exposure to humidity and other weather conditions. Such long term reactions lead to case debonding, propellant cracking, a change in ignition or burning rate characteristics, separation of the plasticizer or

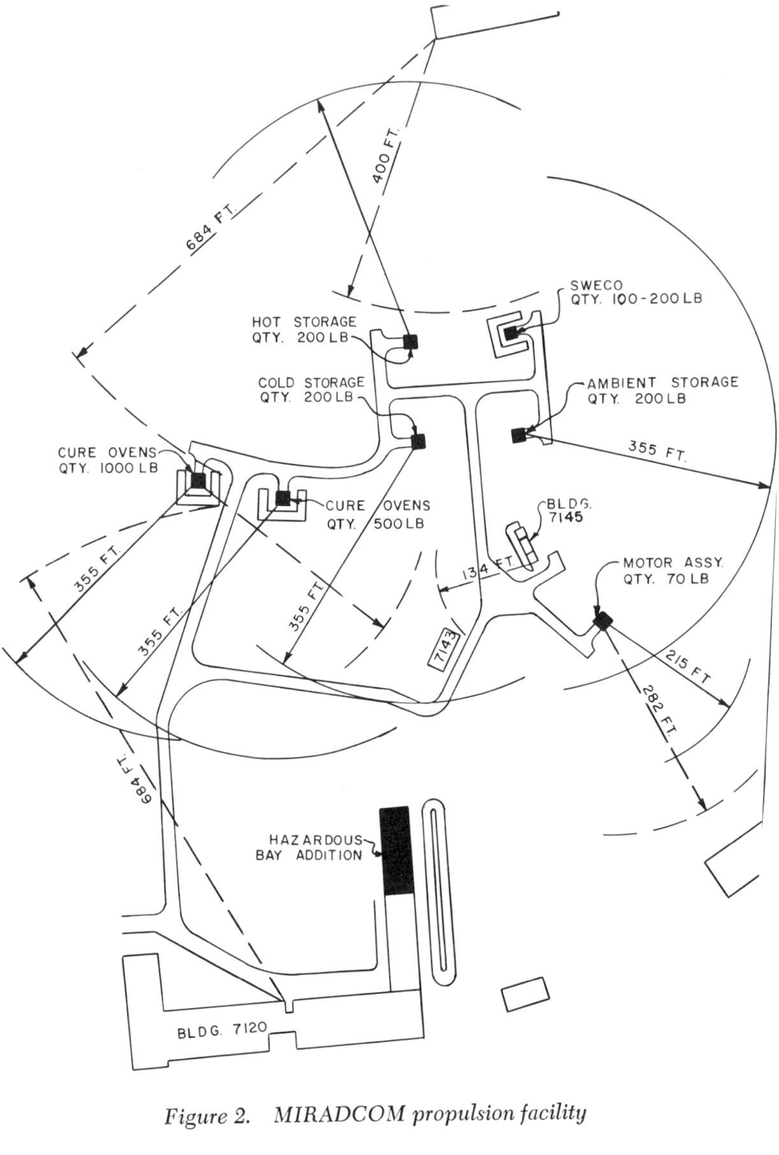

Figure 2. MIRADCOM propulsion facility

a change in physical properties. Such phenomena produces hazards by changing the combustion characteristics which leads to explosions at launch. Investigation of the long term chemistry begin at the ingredient selection stage and continues as long as the missile is in the inventory.

Propellant Mixing. Propellant mixing is the most hazardous operation in propellant development; however, mixer technology is now highly developed. Mixing is always performed remotely. Propellant mixing is scaled from 50 grams/batch mixes for experimental propellants thru standard sizes of one pint, one quart, one gallon, five gallons, fifty gallons, and four hundred and fifty gallons. Propellants that are not fully characterized are almost never mixed in quantities larger than one gallon. Mixing is a process where some friction between ingredients and friction between the propellant and the mixer blades is inevitable. Heat is produced by the stirring process and is removed by water cooling. Mixing is generally done under vacuum to avoid air entrapment into the propellant. Designs to prevent propellant mix from getting into bearings (vertical sigma blade mixers), safety features to prevent mixer blades from contacting mixer bowls (periodically checked for clearance), non-sparking parts, automatic deluge systems (mounted inside as well as outside of large mixers), automatic ingredient feeders, and techniques to remotely cast the propellant into motors are features employed routinely. All equipment is electrically grounded. Process variables such as mixer speed, mix temperature, mix time, and on larger batches mix viscosity are monitored continuously.

Casting. Casting is the final phase of the mixing process. The primary safety consideration in casting is to fill the rocket motor with propellant mix free of voids. Voids produce increased burning surface which can lead to motor case rupture when fired or possibly a transition from deflagration to detonation (DDT). Gas filled voids can produce subsurface ignition from adiabatic compression during ignition load shocks or in vibration testing. Such ignition will lead to over pressurization, motor rupture and possibly detonation. For these reasons most casting as well as mixing is done under a vacuum.

Curing. A propellant mix is similar to any other formulation in that many chemical reactions must occur to reach the final desired "inert" state with the appropriate physical properties. This process of polymerization and cross linking of the binder, bonding to the motor case, and formation of appropriate internal structure is called curing. The cure reactions are done at temperatures from ambient to about 160°F over a period of a few hours to two weeks depending on the propellant and the desired end properties. Oven temperatures are closely controlled and monitored continuously. The safety approach is to avoid contact during the cure process and to conduct the curing in isolated barricaded buildings (see figure 2).

Motor Preparation. After curing, the casting fixtures are removed from the motor. Major items such as removing the core are done remotely. If there are any rough edges on the propellant which require triming the operation is done remotely. The major hazard is fire. Appropriate protective clothing, non-sparking tools, automatic deluge systems, and protective shields are used. Final assembly of such items as igniters and nozzles are relatively routine. Screw thread attachments are usually avoided. Snap rings and other techniques are used. Inspection of small experimental motors is visual. Larger motors and production items are x-ray inspected.

Motor Evaluation. Many different operations are performed in propellant and motor evaluation. The main safety precaution is to keep personnel contact to an absolute minimum. Most operations are performed remotely, the only contact required being during assembly and installation or removal of the test device. The most hazardous operation, test firings, are performed remotely in reinforced concrete barricaded cells (see figure 2). The technician who connects the ignition circuit, the last operation before firing, retains the key to the firing circuit such that it cannot be activated until he returns to the control room.

Hardware Preparation. Motor preparation and motor evaluation result in used hardware to be recycled, along with new hardware, into the propellant and propulsion system development process. Cleaning used hardware includes removal of residual propellant from casting fixtures which can be a serious fire hazard. Appropriate protective gear and deluge systems are employed. Cleaning and degreasing utilize ultrasonic cleaners and degreasing vats which contain toxic fluids such as trichloroethylene in large quantity. Adequate protective equipment and ventilation are required. Coatings, insulators, and liners such as teflon, epoxy and RTV are done routinely with proper precautions for ventilation and protective clothing addressed.

The Missile Research and Development Command (MIRADCOM) Propulsion Facility

Figure 2 shows the MIRADCOM facility including an addition in progress to modernize and consolidate operation into one location. Building 7120 houses the propellant chemistry laboratories. The barricaded north wing houses the hazardous operations cells. These cells are used to perform such operations as experimental propellant mixing thru the five gallon batch size and includes motor loading. These cells are constructed of reinforced concrete of sufficient thickness and strength to contain detonations of quantities up to the limits cited without spalling concrete on the outside walls. Figure 3 shows the arrangement of a segment of the cells. Most operations shown in the work flow process are identifiable in the facility, Figure 2. Loaded motors from the hazard cells are placed in the barricaded cure ovens. The motors are assembled and prepared for tests in the motor assembly building, test fired in bldg 7145 remotely from the control bldg 7143. Conditioned storage is available in the hot, cold and ambient storage buildings. Ammonium perchlorate is ground in the SWECO building. Appropriate safety features considered and followed are summarized in Tables I, II and III. Various distances between buildings are calculated from empirical equations derived from experimental data (1). The distances are based on allowable overpressures. Table IV summarizes equations and typical calculations used to plan this facility. Fragmentation dilutions are derived by direct ratio to experimental data on a 500 lb bomb fragmentation distribution. In Figure 2, distances noted by solid line are fragmentation distances. Other distances are noted by dashed lines.

Safety Criteria for Facility
Table I

1. Perimeter fence with appropriate signs to restrict and limit access

2. Quantity-distances established to keep overpressure below window breakage level for inhabited buildings.

3. Barricades, quantities, distances established to keep lethal fragments below a density of one per 600 sq ft at personnel location.

4. Warning lights and siren to alert personnel of impending tests or hazardous operations

5. Gates to restrict personnel and vehicle flow

6. Lightening protection on each building.

7. Standard Operating Procedure for each operation.

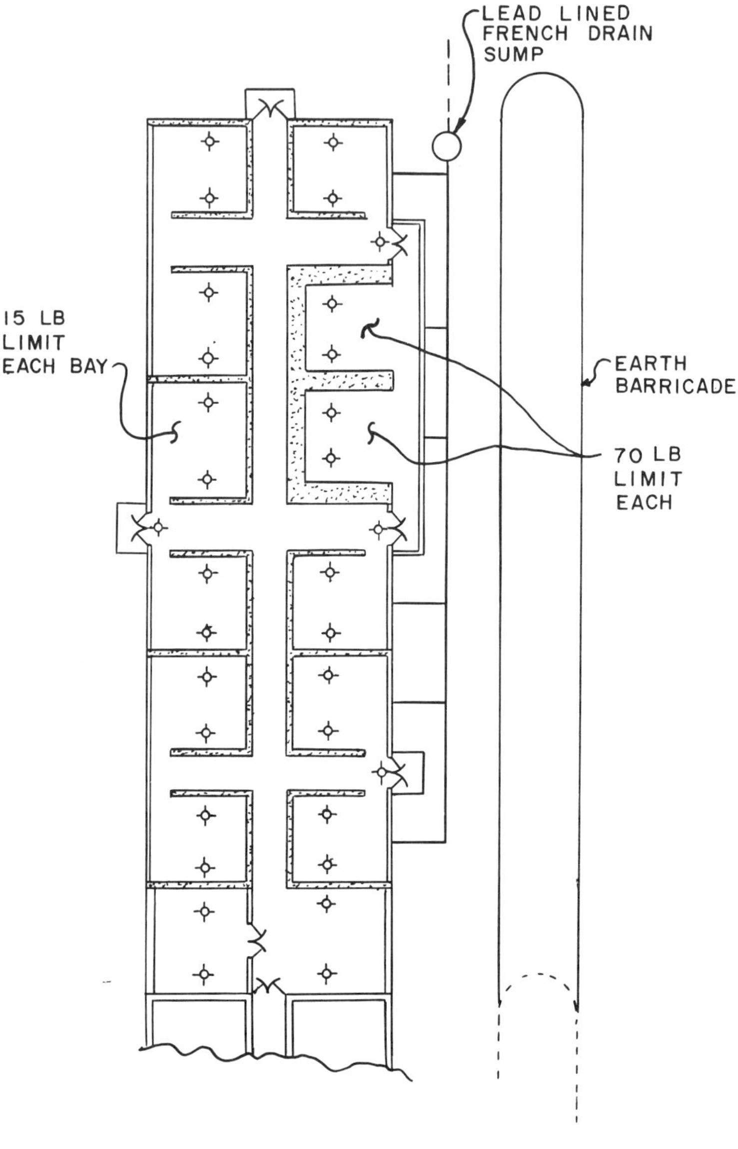

Figure 3. Hazard cell construction

Safety Criteria for Hazard Bays
Table II

1. Twelve inch (12") reinforced concrete walls for 15 lbs explosive limit bays (1 gallon mixer).

2. Thirty six inch (36") reinforced concrete walls for 70 lbs explosive limit bays (5 gallon mixer).

3. Light fixtures, electrical outlets, electrical switches, all electrical connections are explosion proof.

4. Once thru temperature and humidity controlled air.

5. Lead lined drains to sumps.

6. Conductive floors in all propellant handling bays.

7. Foam insulation panel exterior walls for "run thru" exits.

8. Light weight precast concrete "blow-off" roofs.

9. All systems and devices grounded to standard grounds.

10. Standard grounds calibrated annually.

11. Automatic water deluge systems with fast acting sensors and valves.

12. Standard operating procedures cover all processes.

Supporting Safety Organizations and Agencies

Considerable assistance is given by various safety and health organizations. Some of these are internal to MIRADCOM and their counterpart or equivalent exists in any major organization conducting simular operations. Others are more broad based and their service boundary less well define. The following summary is not all inclusive but describe the activities that have major impact on MIRADCOM operations.

MIRADCOM Safety Office. The safety office contains professional safety engineers with considerable experience in dealing with hazardous operations. They maintain up to date knowledge and information on safety equipment and procedures. Files are kept on all accidents and incidents. They review and approve all standard operating procedures and serve as a source of advice and information on all safety matters.

Safety Criteria for Propellant Chemistry Laboratory
Table III

1. Solvents stored in sheds outside of the laboratories.
2. Acids stored in sheds outside of the laboratories.
3. Gas bottles stored in sheds outside of the laboratories.
4. Substances separated into oxidizers, fuels and inerts.
5. Small solvent bottles ($\sim$ 1 liter) used in laboratories.
6. Solvents pumped into containers to avoid vapors and droppage.
7. Hoods exceed OSHA standards ($\geq$ 150 ft/min).
8. Electric heat used - no open flames, matches or smoking permitted in laboratories.
9. No eating or drinking permitted in laboratories.
10. Equipment connected to standard grounds and conductive floors used in propellant handling laboratories.
11. Safety goggles, safety shoes, and flame proof clothing required.
12. Persons never perform laboratory work alone.
13. Detached observer present for all hazardous operations.
14. Standard Operating Procedures cover general laboratory operation and all special processes.
15. Each laboratory equipped with eye wash fountains, manual showers, fire extinquishes and fire blankets.
16. Minimum required material always used.
17. Generally comply with criteria and safety outlined in Sax's Danagerous Properties of Industrial Materials.

Quantity - Distance Calculations
Table IV

	90 lb TNT 70 lb Prop	256 lb TNT 200 lb Prop	640 lb TNT 500 lb Prop	1280 TNT 1000 lb Prop	Formula
Inhabited Bldg					
● Barricaded	178	254	344	434	$R = 40 \sqrt[3]{W}$
● Unbarricaded	359	508	688	868	$R = 80 \sqrt[3]{W}$
No Window Breakage	282	400	543	684	$R = 63 \sqrt[3]{W}$
Public Highway					
● Barricaded	107	152	206	260	$R = 24 \sqrt[3]{W}$
● Unbarricaded	215	304	413	521	$R = 48 \sqrt[3]{W}$
Intraline					
● Barricaded	43	61	82	103	$R = 9.5 \sqrt[3]{W}$
● Unbarricaded	85	121	163	206	$R = 19 \sqrt[3]{W}$

Assume Class 1, Division 1 Propellants of 1.28 TNT Equivalents
R = Distance in Feet W = Weight of TNT in Pounds

Preventive Medicine Activity (MIRADCOM). This operation surveys all processes and procedures for medical safety. They review (and approve in selected instances) standard operating procedures. Laboratories are surveyed for toxic chemical levels. Also, chemical procedures are monitored to determine the level of exposure to toxic substances. They perform annual physical examinations on all personnel involved in hazardous operations or exposed to hazardous chemicals. Medical histories are maintained. Techniques of emergency treatment for both standard and unusual incidents are kept updated.

Army Environmental Hygiene Agency. This Army agency, located at Aberdeen Proving Ground, Maryland, evaluates the toxicity of new chemical compounds. It develops and maintains environmental safety standards. It is an operating arm of and reports to the Surgeon General of the Army.

Department of Defense Explosive Safety Board. This activity reviews and approves all explosives handling facilities operating for the Department of Defense. This includes all industrial and government activities in the United States and selected activities in NATO. They are the final authority on explosives facilities. Their approval was required on the facility described in this paper.

EPA and OSHA. All Department of Defense (DOD) activities are subject to EPA and OSHA regulations including the recent Toxic Substances Control Act. There is a provision for DOD to circumvent requirements for purposes of urgent national defense. However, this provision can only be activated by presidential approval and is unlikely to be utilized except in a national emergency.

Standard Operating Procedures (SOP)

Every operation is covered by a SOP. These procedures delineate the step by step process to be followed in conducting a hazardous operation. They identify and specify the safety equipment and clothing to be employed and the emergency procedures to be followed if an accident occurs. They identify the responsible individual for the operation and specify the number of operating personnel that can be present. The SOP is prepared by the operating personnel, reviewed by the laboratory director, reviewed by and co-approved by the Preventive Medical Activity when health hazards are involved, and approved by the MIRADCOM safety office.

Standard operating procedures serve several purposes. Foremost, they require that the process be thoroughly planned, all known safety procedures be incorporated, and the operation made as independant as possible of personnel changes. In the event of an incident they aid in pinpointing the step responsible for the accident (i.e. it helps identify the unsafe act). In the event of injury or loss of equipment, it serves as a legal document to show that all known precautions were employed. It serves as a teaching aid for new employees.

Summary

This paper describes the MIRADCOM Propulsion Facility Plan, the work flow process, the safety factors that must be considered and the primary organizations that assist in insuring that hazards are kept to an absolute minimum. Safety considerations become a consideration at the initiation of a concept and is a constant partner until the missile is phased out of the inventory. It is only by such constant attention that operations with such materials and devices can be performed without incurring disaster.

Literature Cited

1. Chemical Rockets/Propellant Hazards, Vol I, General Safety Engineering Design Criteria, Chemical Propulsion Information Agency Publication 194, Johns Hopkins Applied Physics Laboratory, Laurel, MD (1971).

RECEIVED November 22, 1978.

6

Prevention of Propellant Flame Propagation through Conveyors Using the Primac/Telemac Sprinkler System

T. W. EWING, F. T. KRISTOFF, and W. T. BOLLETER

Hercules Incorp., Radford Army Ammunition Plant, Radford, VA 24141

As part of the overall Army Modernization Program, a new continuous automated facility is being constructed at Radford Army Ammunition Plant (RAAP) for manufacturing single-base cannon propellants. When completed, this new production plant will be capable of manufacturing 2.5 million pounds per month of single-base propellant. In this plant, new equipment and manufacturing technology have been advanced to produce propellants more economically and with less personnel exposure than in the conventional batch process.

This modularly constructed plant houses the equipment for each major processing operation within bays segregated by twelve-inch thick, steel-reinforced concrete walls. However, the continuous processing concept requires that operations be interconnected with vibratory conveyors passing through holes in the fire walls, see Figure 1. The conveyor design includes a nitrogen purge and oxygen level monitor to preclude hazardous vapor/air mixtures where solvent-wet in-process material is being conveyed. Fast-acting fire gates were initially designed to prevent propagation of fire between adjacent operating bays. As a cost savings measure, alternatives to the fire gates were considered.

A study was initiated to determine if the existing high-speed sprinkler system in the conveyors could detect and quench a fire and prevent flame propagation between bays. The effectiveness of this system was determined by establishing the minimum number of sprinklers and water needed to prevent a dry M1SP, 105mm propellant fire from propagating through a typical, full-scale conveyor. A concurrent study developed a flexible seal to prevent flame from propagating through fire wall openings around the outside of the conveyor.

Equipment and process design avoided process conditions which could result in an explosive reaction by preventing unnecessary and potentially explosive accumulations of in-process and finished materials and offering minimum confinement to the propellant by incorporating pressure relief venting.

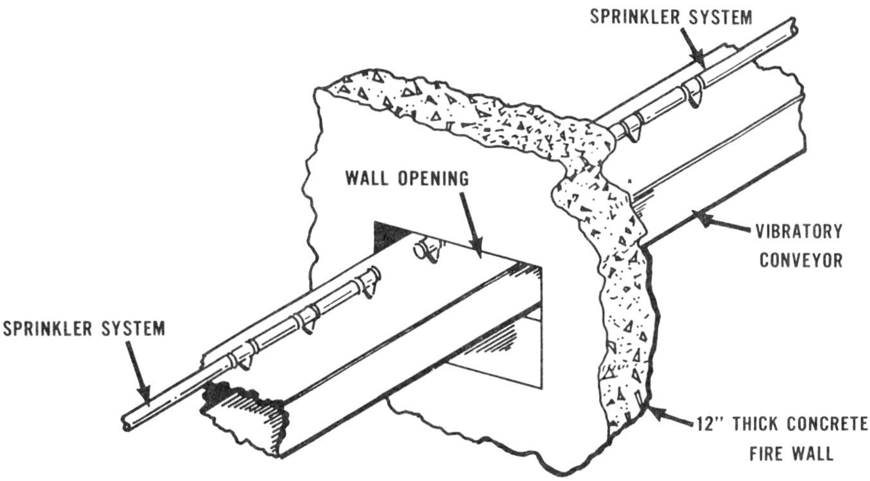

Figure 1. Typical conveyor/fire wall arrangement

DISCUSSION

Process Description

Figure 2 shows the major manufacturing operations in the continuous automated single-base line (CASBL).

The process begins by pumping a 90/10 water/nitrocellulose (NC) slurry to a thermal dehydration unit. The NC is dried, sprayed with alcohol and conveyed to the compounding operation via a vibratory conveyor. The compounder combines the NC with other propellant ingredients, ether and additional alcohol to form a 26 percent solvent-wet paste. Other propellant ingredients include dinitrotoluene (DNT) and potassium sulfate. The paste is conveyed to a rotating screw type mixer for further solvent addition, additional mixing and consolidation into pellets. The pellets are conveyed to a vertical screw extruder and extruded into solvent-wet perforated strands. The strands are conveyed via water in tubes to the strand cutter and cut into short, uniform lengths (granules). The solvent-wet granules are mixed with water and pumped to the first of three drying operations. Most of the free solvent is removed from the granules by hot air in the solvent recovery operation. The remaining solvent is extracted by hot water sprayed over the granules. Drying of the water-wet granules is completed using fluidized bed type hot air driers. After drying, the granules are weighed into drums and transported to a storage area. During the drying operations, vibratory conveyors are used to convey solvent-wet, water-wet and dry propellant granules from one operation to another. Dry propellant operations are potentially the greatest fire hazard because they ignite more easily and burn more vigorously than solvent-wet or water-wet propellants.

Sustained Burning Tests

It was anticipated that the solvent-wet material in a nitrogen purged atmosphere and water-wet materials would not sustain a burning reaction, or would burn at a comparatively slow rate and be quenched by the sprinkler system with little difficulty. The absence of applicable data required that tests be conducted to affirm which materials would sustain a burning reaction under processing conditions. These tests were conducted by placing a one-inch deep layer of water-wet or solvent-wet propellant or propellant ingredient within a twelve-inch wide closed conveyor up to thirteen feet in length. When solvent-wet propellant was tested, the conveyor was purged using nitrogen gas until the oxygen level was below the lower explosive level. A single perforated propellant granulation (M1SP, 105mm) was selected for all subsequent tests because it is the fastest burning propellant planned to be manufactured in the CASBL.

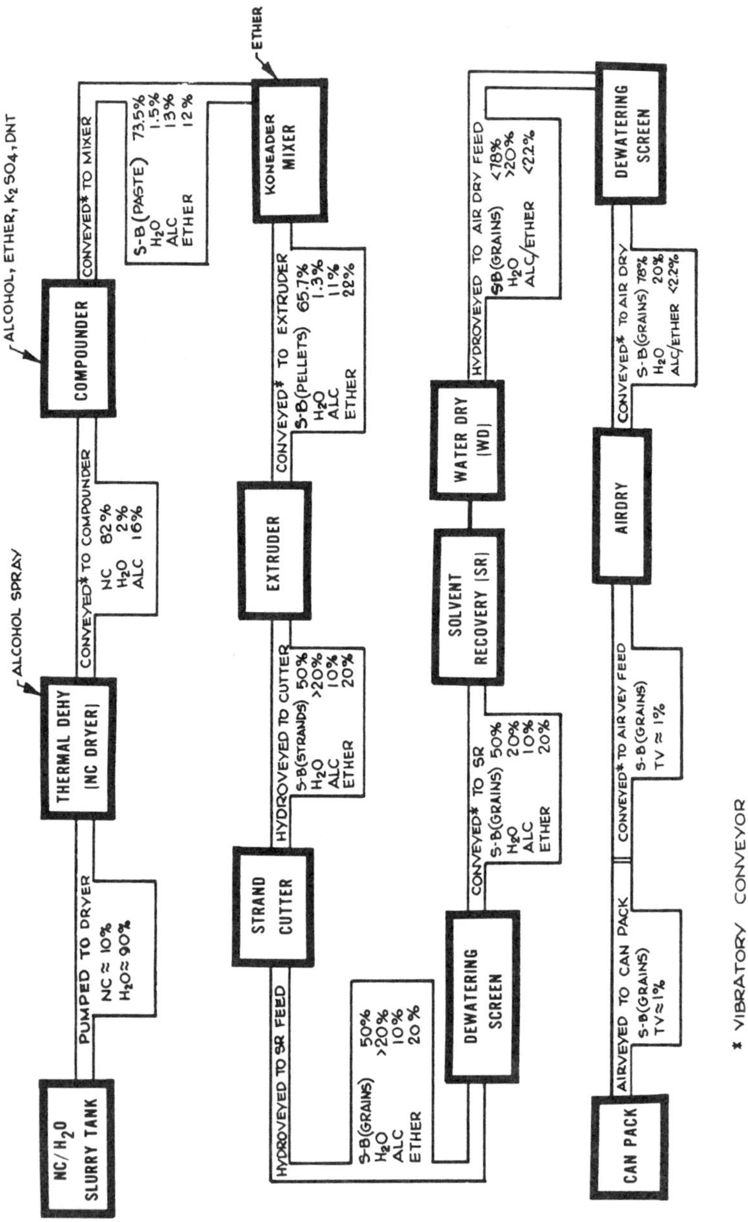

Figure 2. Continuous automated single-base line process schematic. Note: all compositions shown are nominal anticipated at each processing stage.

One end of the propellant bed was ignited with an Atlas match. After burning was complete, the propellant was examined visually to determine if burning propagated the entire length of the propellant bed (sustained burning) or self-extinguished (no sustained burning). The results of these tests are summarized in Table I and show that sustained burning will not occur for dry DNT; ≳ 23 percent alcohol-wet NC; ≳ 12 percent water-wet (surface wet) M1SP granules in a closed conveyor; or ≳ 14 percent alcohol/ether-wet M1SP granules. Additional tests determined that ≳ 18 percent water-wet (surface wet) granules in an open conveyor will sustain a burning reaction.

Table I - Summary of Sustained Burning Test Results

Material	Physical State	Sustained Burning Threshold, % 1/
DNT	Dry, Ground	Will not burn in conveyor
NC	Alcohol Wet 2/	23
M1SP Finished Granules	H_2O (Surface) Wet	> 18 (open conveyor) 12 (closed conveyor)
M1SP Green Granules	Alcohol/Ether Wet 2/	14

1/ Sustained burning at next lower test level

2/ Nitrogen purged atmosphere during tests

Application of these data to CASBL show that most water and/or solvent-wet propellant and propellant ingredients will not sustain a burning reaction for the conditions which will exist in CASBL (Figure 2). These tests demonstrated that the existing conveyor sprinkler system is more than adequate to prevent between bays flame propagation in most CASBL locations.

Minimum Water to Quench Fires

As anticipated, sustained burning tests showed that dry M1SP, 105mm propellant (granules) is the most energetic burning material within CASBL conveyors. A typical CASBL conveyor configuration for handling dry propellant is shown in Figure 3. In this operation, dried propellant is discharged from dryers into a feed hopper. The propellant is fed at a uniform rate from the hopper to a vibratory feed conveyor and then to a vibratory collection conveyor. The collection conveyor moves the propellant through a fire wall and to the next operation. Flame quench tests were conducted by igniting a uniform one-inch deep layer

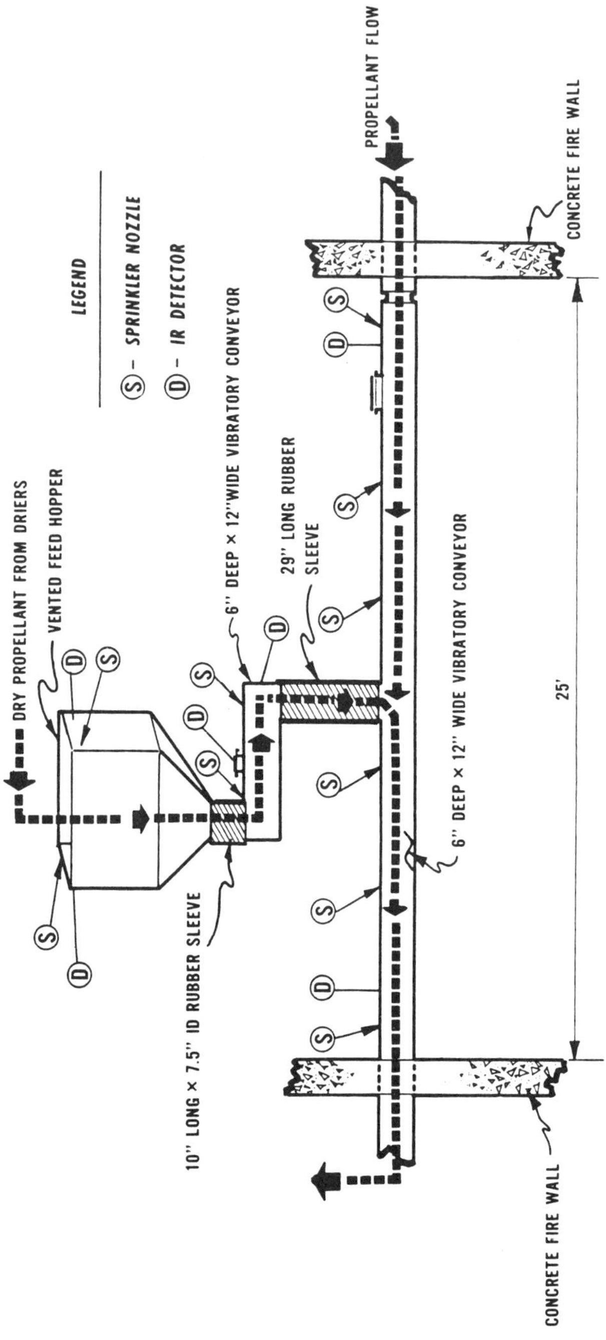

Figure 3. Typical dry propellant conveyor configuration

of M1SP propellant within a full-scale simulated CASBL conveyor. The minimum sprinkler/flame detector arrangement and sprinkler water necessary to prevent flame propagation were determined.

The setup used for these tests is shown in Figures 4 and 5. The test conveyor was thirteen feet long and open at each end. A four-foot long section of conveyor was connected to the ignition end of the test conveyor by a rubber boot to simulate vibratory feeder connectors normally found on CASBL conveyors. Four sprinkler nozzles were welded at three-foot intervals into the conveyor lid and canted 45 degrees toward the ignition end of the conveyor. In tests requiring fewer than four nozzles, one or more nozzles were blocked using pipe nipples. Water pressure was varied between 25 and 90 psig using a pressure regulator in the supply line. One infrared detector was used to detect propellant burning and activate the sprinkler system via a Primac valve. [Infrared (IR) detectors manufactured by ADT for use with the Primac valve are manufactured and sold by Grinnel (Telemac Ultra-High Speed Fire Detection System). The Primac valve consists of two primer caps activated electrically when an IR detector responds to a fire.] This detector was located approximately two feet in front of the first active sprinkler nozzle for all but confirmatory tests. Ignition of the propellant was accomplished remotely using two Atlas matches located at the bottom of the conveyor (submerged ignition).

Infrared detectors were also used to detect the presence of flame at various conveyor locations including the end of the conveyor. High-speed color movies (200 frames per second) monitored each test to verify that flame propagation did, or did not, occur at the end of the conveyor. Rupture vents, determined to have little effect upon test results in preliminary trials, were blanked off using steel plates for some of the quench tests.

Testing was conducted in two stages. Stage one involved testing a one-inch deep uniform layer of propellant and simulates one-inch deep propellant layers expected during normal operations. Stage two tests were conducted using a one-inch deep propellant layer in most of the conveyor , and a six-inch deep by twelve-inch long layer in the ignition end of the conveyor. Stage two tests simulated the worst case involving dry propellant. The results of each test stage are discussed separately below.

Stage One - Quench Tests

Twenty-five quench tests were conducted with a one-inch layer of M1SP, 105mm propellant in the simulated CASBL conveyor. Approximately forty-eight pounds of M1SP was used during each test. Pressure relief vents were blanked off. As can be seen from test results in Table II and summarized in Figure 6, a single water nozzle spraying 17 gpm of water at $\gtrsim$ 55 psig pressure at any of the first three nozzle locations tested will

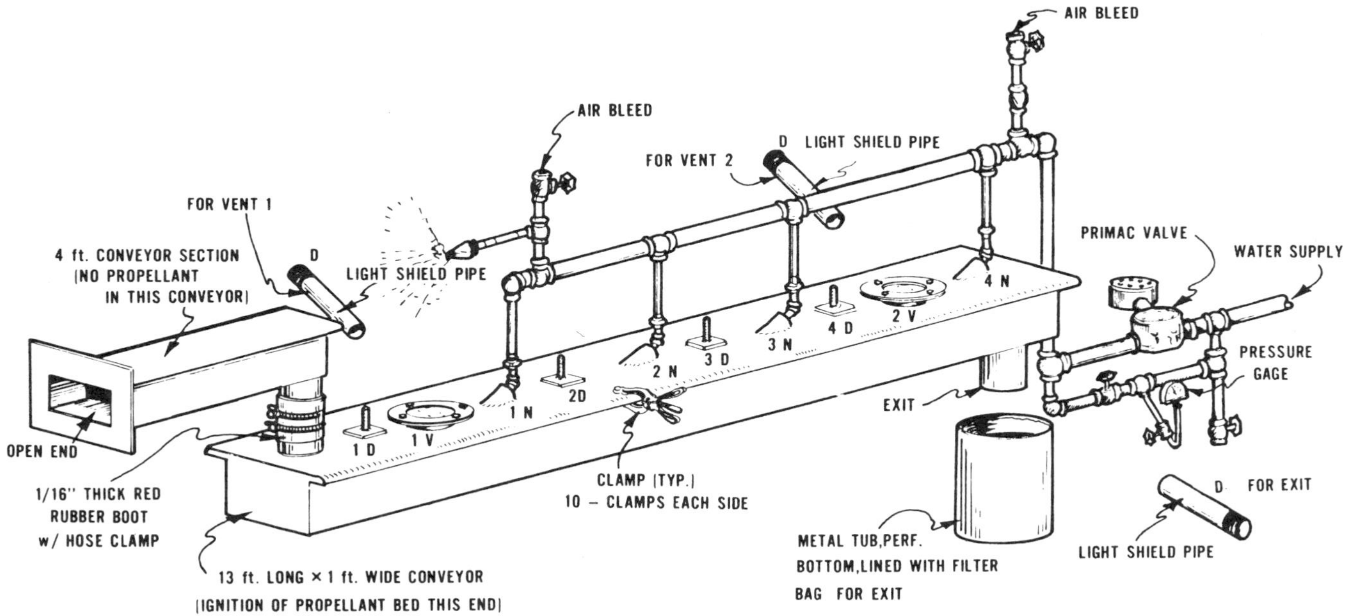

Figure 4. Full-scale conveyor test setup. (D) ID dector, (N) springler nozzle, (V) pressure relief vent.

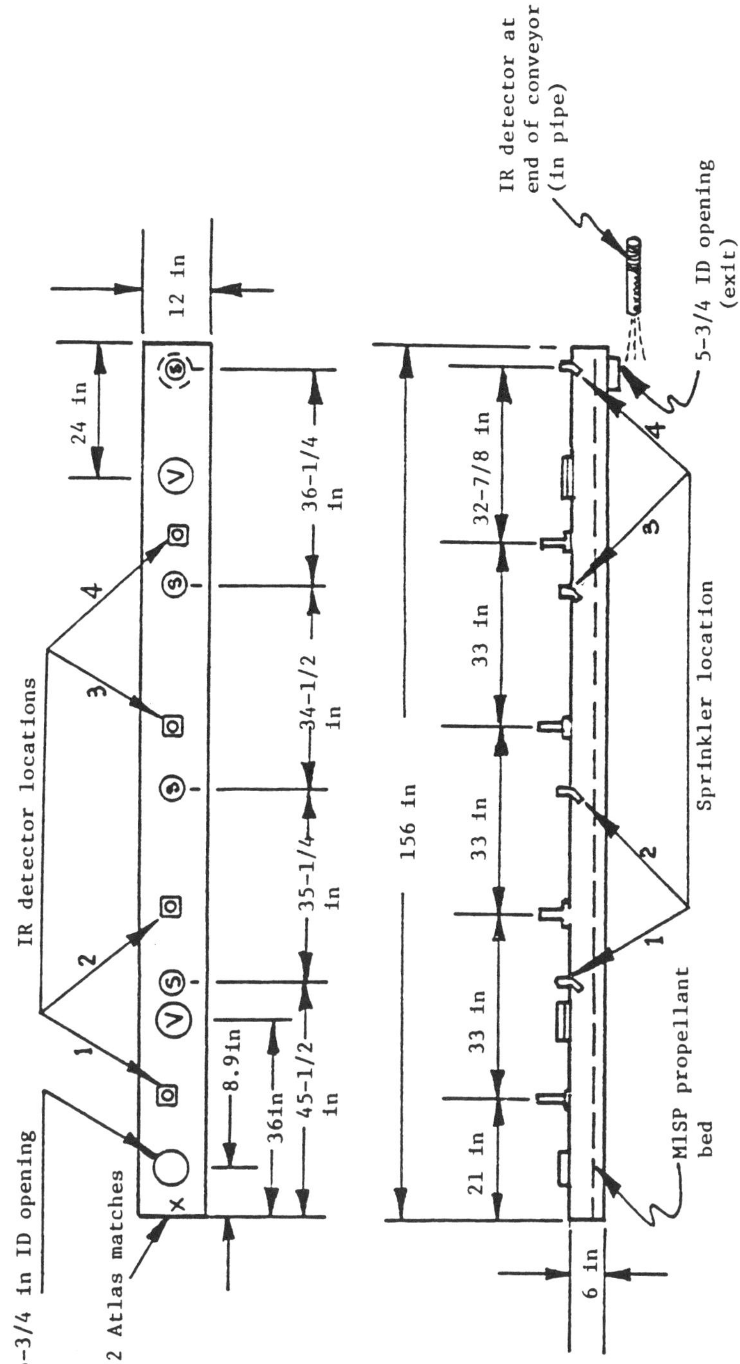

Figure 5. Simulated CASBL conveyor. (V) Locations of 6 in ID conveyor pressure relief vents (when used); (S) locations of sprinkler nozzle fittings welded to top of conveyor.

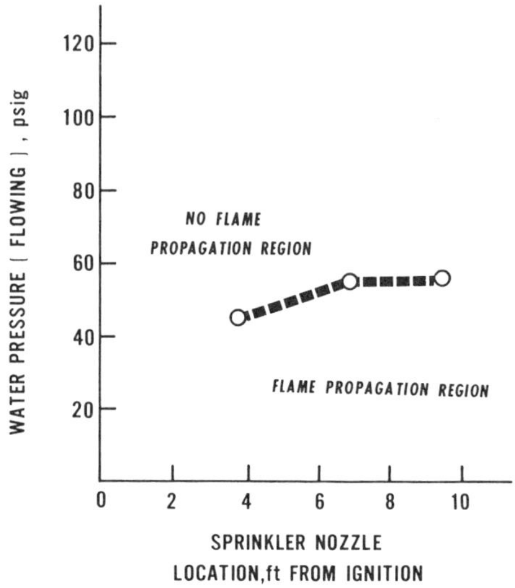

Figure 6. Minimum quench requirements. One-inch bed dry M1SP, 105 mm.

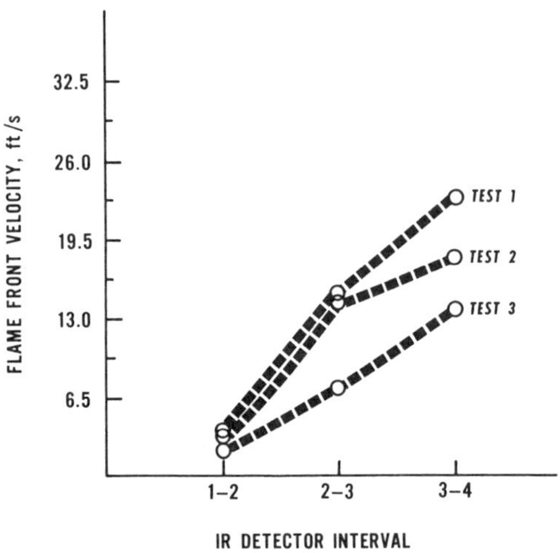

Figure 7. Flame velocities in CASBL conveyors. Test conditions: dry M1SP, 105 mm; 1-in. bed depth; no sprinklers, no vents; ends open.

effectively quench burning M1SP and prevent flame propagation in the test configuration.

Table II - Minimum Water to Quench Burning One-Inch Layer of Single-Base Propellant in a Conveyor

Sprinkler Location	Threshold* Water Pressure psig	Detector Location	Flame Propagation Stopped	No. of Trials
1	45	# 1	Yes	8
2	55	# 2	Yes	7
3	55	# 3	Yes	10

*Defined as minimum H_2O pressure required to quench a fire. Water pressure tested below this value resulted in flame propagation. A minimum of three trials were conducted at indicated pressure.

Table II data also show that a slightly greater water pressure is needed for quench as the nozzle is moved further away from the ignition location. This difference in water requirements is apparently caused by increased flame velocity achieved during longer run-up distances to the sprinkler nozzles. Increased flame velocity represents a more severe burning reaction and is more difficult to quench. Figure 7 shows the average flame velocities between detectors for three trials conducted without sprinklers. These data show that burning velocity increases as burning progresses down the conveyor and indicate that early flame detection and rapid sprinkler activation are necessary to prevent flame propagation in conveyors.

Stage Two - Quench Tests

Tests conducted with the propellant build-up configuration in the ignition end of the conveyor resulted in more erratic and aggressive burning reactions than observed in Stage one tests. Approximately 60 pounds of M1SP was used for each test. For these tests, aluminum rupture discs 0.0008-inch thick were used on each pressure relief vent. One IR detector activating the sprinkler system was located in No. 1 (see Figure 4) position.

As can be seen from data in Table III, a simultaneous activation of two water nozzles spaced six feet apart and spraying water at 90 psig is required to quench and prevent propagation of flame through the conveyor for the propellant build-up condition. The two water nozzles operated at 90 psig provide a water flow of 21 gpm (each nozzle) which is a minimum value for this severe process situation.

Table III - Minimum Water to Quench Built-Up Single-Base Propellant Layer in a Conveyor

	Sprinkler Location	Threshold* Water Pressure psig	Flame Propagation Stopped	No. of Trials
Single Nozzle Tests	1,2 or 3	$>$105	No	4
Multiple Nozzle Tests	1,2,3 & 4	$\overline{<}$ 95	Yes	1
	1 & 3	90	Yes	7
	1,2 & 3	$>$75	No	2

*Defined as minimum H_2O pressure needed to quench a fire. Water pressures tested below this value resulted in flame propagation.

Single water nozzles located in conveyor positions 1, 2 or position number 3 and spraying water at 105 psig failed to quench the fire and prevent flame propagation. Also, multiple water nozzle tests involving simultaneous activation of three or four nozzle combinations were only partially successful when lower water pressures were tested. It is noted that the three-water-nozzle test combination operated at 75 psig delivered 39 percent more water in terms of gpm than the two-water-nozzle test at 90 psig. Therefore, water pressure, number of nozzles and nozzle location are important factors in preventing flame propagation through the conveyor.

Confirmatory Tests

Seven confirmatory tests were conducted using the modified test setup shown in Figure 8. These tests were conducted to assure that additional propellant within a conveyor feed hopper would not result in a more severe burning condition than previously tested. Three nozzles with 90 psig water pressure were used for these tests since there were at least three nozzles on actual CASBL conveyors. One-hundred ten pounds of M1SP propellant was used for each test. Initial tests used two IR detectors in the upper conveyor. Table IV shows that two of four tests stopped flame propagation.

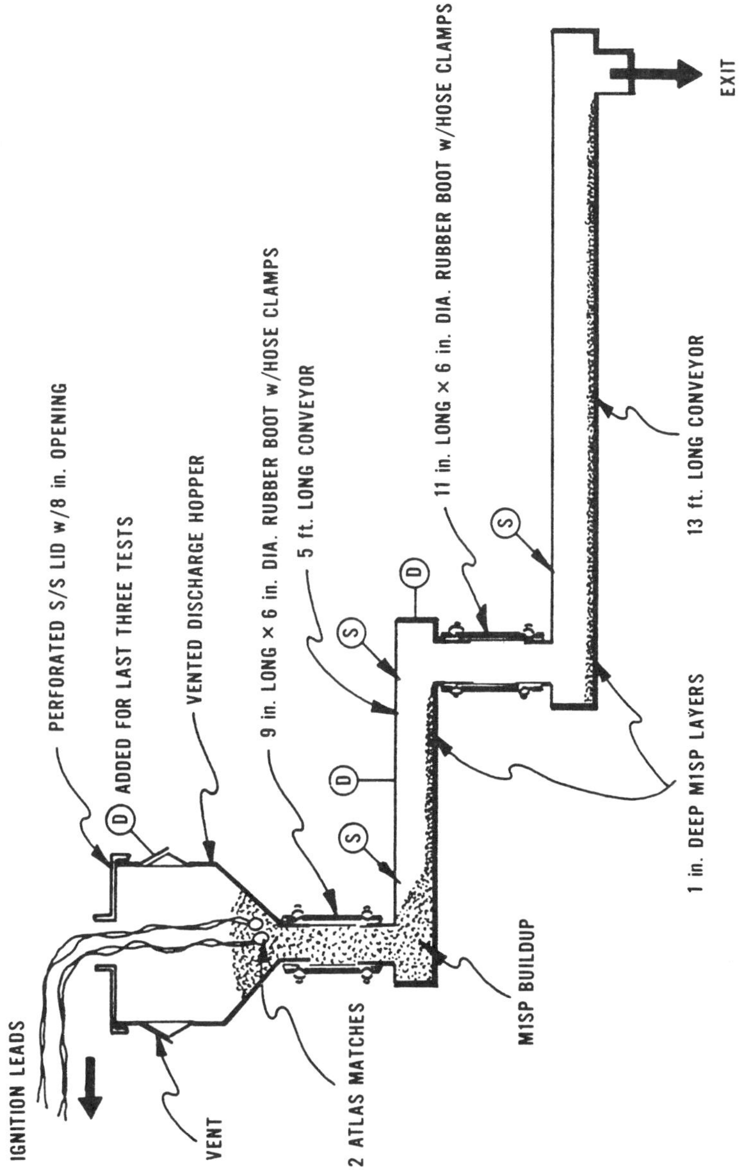

Figure 8. Confirmatory flame quench test setup. (S) *sprinkler nozzle location,* (D) *telemac IR detector location,* (∴) *dry M1SP, 105 mm propellant.*

Table IV - Confirmatory Flame Quench Trials

Test Conditions	Water Pressure psig	Flame Propagation Stopped	No. of Trials
3 Nozzles; IR Detectors in Conveyor only	90	Yes	2
	90	No	2
3 Nozzles; IR Detectors in Conveyor and Hopper	90	Yes	3

A review of the data and high-speed movies taken of the tests indicated that smoke may have prevented IR detectors from seeing the fire soon enough to prevent flame propagation. A detector was added within the hopper, and this sprinkler system arrangement effectively prevented propagation in three consecutive trials.

CONCLUSIONS

For most dry M1SP propellant (one-inch bed depth) and other wet and dry in-process propellant states, a single water nozzle spraying 17 gpm of water at 55 psig pressure is sufficient to prevent flame propagation in conveyors.

The propellant build-up condition presents more severe burning. Therefore, two water nozzles spaced six feet apart and spraying 21 gpm (each) of water at 90 psig pressure are required to prevent flame propagation in conveyors.

Adaptation of these test results successfully demonstrated effective detector response and fire quench in confirmatory tests simulating a feed hopper/conveyor arrangement.

These study findings were employed to optimize Primac/Telemac sprinkler systems in all CASBL vibratory conveyors. Design optimization studies provided a fire detection system employing minimum water which will effectively prevent between-bay flame propagation in the event of a process fire. The cost of installing quick-acting mechanical fire gates was avoided without affecting system safety of the CASBL.

ABSTRACT

Safety testing established the minimum Primac/Telemac sprinkler system design and operating requirements for preventing a fire from propagating through a bed of granulated cannon propellants in conveyors. Simulated test configurations involved several propellant thicknesses. Test results are applicable for assessing safety and optimizing sprinkler designs in manufacturing operations involving single-base cannon propellants. Application of these test results to the design of a modern continuous automated cannon propellant manufacturing facility at Radford Army Ammunition Plant resulted in the substitution of Primac/Telemac sprinkler systems for quick-acting mechanical fire gates and optimizing the Primac/Telemac sprinkler design to achieve a high level of system reliability for preventing flame propagation between adjacent bays via conveyors.

RECEIVED November 22, 1978.

Design Criteria for Mobile Ammunition Surveillance Shop Including Personnel Protection Consideration

R. N. HUDDLESTON

DARCOM Ammunition Center, SARAC-DEN, Savanna, IL 61074

Materiel readiness is a primary consideration and a key element of National Defense. Consequently, the timely detection and correction of deficiencies is essential to assure material readiness and to preclude the corresponding economic loss that would accrue should deficiencies remain undetected.

Supply Bulletin 742-1 describes the type of inspections performed and procedures to be utilized in the surveillance of ammunition. Review of the variety of ammunition storage locations has indicated that in overseas areas many of the storage sites are remote from base depots and a surveillance workshop (Figure 1). Due to the lack of facilities, inspections were being accomplished by transporting samples to be tested from storage to the depot and then returning them to the storage site. In some cases the inspections could be performed at the storage location but necessitated the movement of appropriate tools and equipment to the site by whatever means was available and on many occasions required the use of privately owned vehicles. This procedure resulted in an uneconomical operation at best. The transportation of the ammunition samples from the storage site to the base depot and return through civilian domain also resulted in additional liability potential should an incident occur as well as exposing the ammunition to an added security risk from dissident groups.

TASK:

To resolve these problems, the DARCOM Ammunition Center was tasked to design a Mobile Surveillance Inspection Shop that would provide a suitable work location and contain all of the necessary tools and equipment to accommodate the variety of inspections when conducted at the storage site. This mobile shop was to be transportable and selfsustaining. It would be designed to insure ease of operation, service and storage of the unit, as well as to provide an efficient work area with a maximum consideration for the protection of personnel exposed to both the industrial and explosive hazards inherent to the inspection function. Safety criteria associated with this design includes all of the requirements applicable to a fixed facility as well as those peculiar to the required mobility of the unit.

SAFETY GUIDANCE:

Briefly, explosive safety consists of minimizing and/or eliminating conditions which might lead to the accidental detonation of explosives and/or providing necessary protection to operating personnel that would minimize and/or eliminate personal injury associated with an explosive incident.

Army experiences as well as technological advances have resulted in the accumulation of substantial data, which has been used to develop the functional criteria, that must be recognized in design and development of equipment and associated facilities used for ammunition operations. Primary guidelines are found in the Army Materiel Development and Readiness Command Safety Manual (DARCOMR 385-100), Ammunition and Explosive Standards (TM 9-1300-206) and Occupational Safety and Health Act Standards, as well as other laws, rules and regulations. All of these criteria have been recognized and considered in the development of the Mobile Surveillance Inspection Shop.

APPLICATION OF DESIGN CRITERIA:

A variety of common and basic transportation units were studied to determine which unit could best be utilized to house the Mobile Surveillance Inspection Shop. The Army MILVAN container (Figure 2), was determined to be the most capable of being adapted for the intended purpose and to provide the greatest benefit. Additionally, the capability to transport and handle the basic MILVAN unit is universal at all installations within the Army.

The majority of surveillance operations planned to be accommodated in the mobile shop involve items where the explosives are contained within projectiles, cartridge cases, fuzes, grenades and are not normally exposed to conditions that would be considered as being conducive to an explosive incident. Due to the existence of the explosive potential, however, facilities design and safety considerations as well as internal features and equipment were developed to criteria normally associated with exposed explosive conditions. Electrical wiring, controls, lighting and other electrical devices inside the mobile shop (Figure 3) were housed in enclosures selected to meet the requirement for hazardous locations as advised in the National Electronics Code for Class 1, Group D and Class 2, Groups E, F, and G conditions. These precautions included the thermostat utilized to control the temperature inside the shop.

Compressed air capability was piped to a convienient location

Figure 1. Ammunition surveillance being performed under field condition at an Ammunition Supply Point (ASP)

Figure 2. U.S. Army MILVAN's, at Transportation Testing Area, DARCOM Ammunition Center, Savanna, IL

Figure 3. Interior view of Mobile Surveillance Inspection Shop showing explosion proof electrical switches mounted on forward bulkhead, explosion proof light enclosures, office and grenade disposal barricade

inside the facility work area to provide a non-sparking power source for power tools and equipment.

Auxilliary and support equipment, required to sustain the operation and not easily adapted to use in the explosive environment, was located in an equipment module (Figure 4) attached to the end of the shop and sufficiently isolated to preclude any hazard to operations conducted therein. The gasoline driven electric generator supporting total electric power requirement, a heat pump, air compressor and hydraulic pump are all located within this equipment module. A sectionized rod with self-contained hammer is provided and stored in the equipment module. This rod is driven into the ground at each work location and connected to the shop frame to insure electrical grounding and prevention of any buildup of static electricity that could produce a spark within the shop.

This basic MILVAN unit has been equipped with four hydraulically operated legs (Figure 5) that can be extended from pockets on the sides of the mobile shop. The legs were provided as a self-contained means of loading and unloading the shop from the MILVAN chassis or flatbed trailer utilized to transport the item. During field trials, the capability of setting the shop at ground level as compared to when it was mounted on a chassis or flatbed trailer, was proven to eliminate many of the hazards associated with the movement of ammunition in and out of the work location.

WORK AREA LAYOUT:

The Army MILVAN has double cargo doors in one end which are secured with camlock hardware. The locking mechanism was altered to provide a panic bar (Figure 6) to permit rapid evacuation should it become necessary. A pedestrian door with panic hardware was installed in the side of the shop permitting an unobstructed escape route from the opposite end of the operating area.

ARMED GRENADE DISPOSAL:

One of the most potentially hazardous operations in the Shop is the surveillance inspections performed on explosive type hand grenades. The possibility that the safety pin can become disengaged, the handle released and the explosive train initiated to fire the grenade in 5 to 7 seconds, represents a significant hazard. Safety Criteria requires a barricade be provided for ammunition operators to dispose of grenades that become armed or are suspected of being armed. The Ammunition Peculiar Equipment Program designs and provides such equipment for ammunition operations and has a standard "pitch-in" barricade designed for this purpose. The basic design of the standard items, however, was too large for the space available. Additionally, the design was such that pressures and hot gases generated through a grenade detonation could not be harmlessly dissipated in the relatively

Figure 4. Equipment compartment with doors open. Electrical generator is in the upper right-hand corner. Heat pump is upper left and air compressor in the lower left. The three ground rods, driver, and cable are in the center with fire hazard signs storage in lower right.

Figure 5. Mobile Shop at ground level with hydraulically operated lifting legs extended to operating position

Figure 6. Interior view of rear exit door showing panic hardware. Pitch-in-barricade for disposing of suspect grenades is shown to right of exit door.

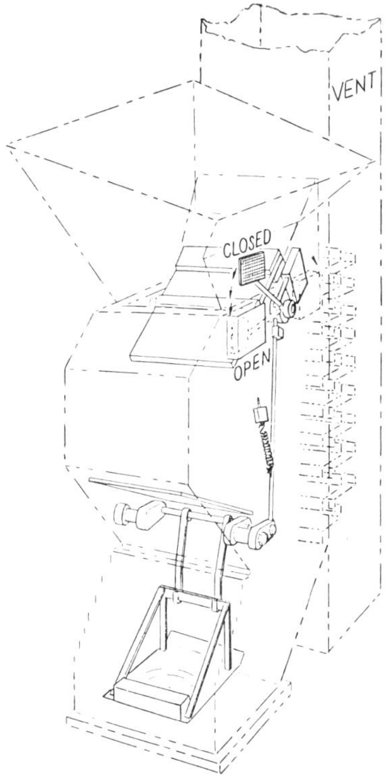

Figure 7. Phantom view of pitch-in-barricade showing operation of blast door and vent

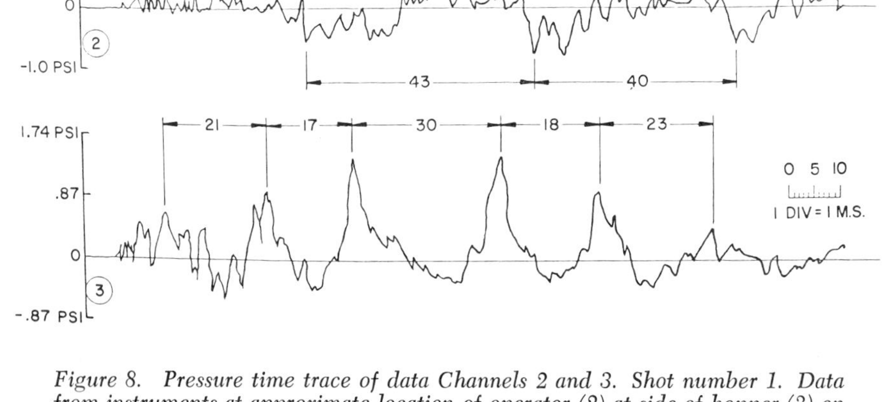

Figure 8. Pressure time trace of data Channels 2 and 3. Shot number 1. Data from instruments at approximate location of operator (2) at side of hopper (3) on pitch-in side of hopper.

Figure 9. Mobile Surveillance Inspection Shop mounted on MILVAN chassis ready for movement

limited volume of the Mobile Shop. To provide an acceptable barricade, meeting this criteria dictated development of a new design. Because the barricade is classified as an operational shield, its design had to meet the criteria in Military Standard 398. A prototype had to be tested in accordance with procedures in this standard to insure compliance with the criteria, which requires that operating personnel not receive a heat flux exposure greater than $0.62t^{-0.7423}$ $cal/cm^2/sec$, where "t" is time of exposure in seconds, or be subject to more than 2.3 pounds per square inch peak positive incident pressure, and that all fragments must be contained.

The acceptable "pitch-in barricade", developed for the Shop, contained a blast door that closes automatically when a grenade is tossed into the opening (Figure 7). Supressive shielding principles were used to control fragments while venting the hot gases to the outside of the Shop. When the prototype barricade was tested with a live grenade, under conditions simulating the Mobile Shop, the blast pressures inside the Shop, adjacent to the barricade, were well within the 2.3 pounds per square inch permitted by the Military Standard criteria (Figure 8). The heat flux was insignificant and could not be measured with the instruments normally used during this type of testing. No fragments from the exploding grenade reached the outside of the barricade. In short, an operator inside the Mobile Shop, using the barricade to dispose of an armed grenade, would not be injured.

FIRE SYMBOLS:

On each of the four sides of the Shop, a holder was provided so that the appropriate fire hazard symbol, reflecting the most hazardous material in the Shop, could be posted. Fire symobls for each of the classes of ammunition that would be encountered in performing field surveillance operations are provided and stored in the equipment module.

SUMMARY:

The availability of the Mobile Ammunition Surveillance Inspection Shop will substantially benefit the inspection function through a more effective capability to determine ammunition serviceability at remote storage locations as well as accomplishing this inspection function at a substantially reduced cost. A greater degree of safety will be afforded to the on-site operation. Exposures of civilian population inhabiting the area through which munitions were transported to a base depot activity will be eliminated. Ammunition packaging can be restored to its original configuration through the ready availability of equipment contained within the Mobile Shop. The availability of this Shop will substantially increase the confidence level associated with an on-site inspection of the ammunition commodity. This Mobile Shop as fielded (Figure 9) provides a fully equipped, safe work environment that can be easily transported to remote sites to fulfill the ammunition surveillance mission responsibility as an adjunct to the overall materiel readiness objective of the Army.

Literature Cited:

1. Ammunition Surveillance Procedures, SB 742-1, Hq. Department of the Army, Washington, DC, May 1976.

2. Safety Manual, AMCR 385-100, Hq. US Army Materiel Command, Alexandria, VA, Latest Edition.

3. Ammunition and Explosive Standards, TM 9-1300-206, Hq. Department of the Army, Washington, DC, June 1977.

4. Shields, Operational for Ammunition Operations, Criteria for Design of, and Tests for Acceptance, MIL STD 398, US Government Printing Office, Washington, DC, 5 November 1976.

RECEIVED November 22, 1978.

8

Safety Design Criteria for the BALL POWDER Process

BILL CARROLL and JOHN J. NICHOLS

Olin Corp., P.O. Box 222, St. Marks, FL 32355

The explosives industry offers a great challenge to the safety conscious designer. The general safety rules and regulations utilized by the explosives industry for basic design criteria are also used as the guidelines for design in the BALL POWDER process. However, many of the traditional process procedures of the explosives industry have not been retained or have been modified for the BALL POWDER process. The manufacture of BALL POWDER involves hazards, with added hazard due to the explosive nature of the material processed. Pumps, pipes, tanks, equipment, etc. are veritable bombs if handled improperly. As a result, there are certain safety design requirements unique to the process.

The intention of this paper is to describe some of the more unusual safety design requirements. In order to better understand these requirements, a brief history of the type of process utilized in the manufacture of BALL POWDER is presented. Next those safety advantages internal to the process are summarized. Specific requirements in the areas of grain formation, nitroglycerine manufacture and transfer, and continuous drying are discussed. Finally, the basic fire protection system utilized is described.

One of the most significant developments in the manufacture of any gunpowder in the last 60 years has been the development of the BALL POWDER process. BALL POWDER was originally produced in a batch method, but an alternate method was developed to overcome the problem of difference in particle size distribution required versus product distribution produced. The new process involved mechanical graining and batch hardening. While this overcame the size distribution problems, the capital cost was much greater than a regular batch plant. Development work was then directed to reducing the investment. The process developed consisted of continuous mechanical graining and continuous hardening. The continuous process exhibited superiority over conventional methods in both performance and process advantages. Increased safety was combined with rapid means of stabilization,

0-8412-0481-0/79/47-096-171$05.00/0

grain formation, and solvent removal.

The basic raw material in the production of BALL POWDER is nitrocellulose. Two different forms of nitrocellulose can be used, virgin nitrocellulose or FNH (flashless, non-hygroscopic, single-base, extruded powder). The dibutylphthalate and dinitrotoluenes in the FNH grain must be extracted before being used in the BALL POWDER process. The first stage in the manufacture of BALL POWDER is grain formation. This operation includes four sequential operations, consisting of (1) lacquer preparation, (2) mechanical graining, (3) grain shaping, and (4) grain hardening. Lacquer preparation is the operation which converts the base nitrocellulose stock to a viscous dough-like lacquer. This is accomplished by the addition of ethyl acetate and various additives to the nitrocellulose in a mixing vessel where intimate mixing produces the lacquer. In the graining operation, the lacquer is converted from a "doughy" mass to individual lacquer particles by extrusion through a mechanical pelletizer. Shaping is performed by transferring a slurry of lacquer particles through a series of jacketed tubes. The combination of liquor (transfer fluid) flow, liquor composition, temperature, and slurry velocity in the shapers causes the lacquer to round up into balls and give up its internal water. The lacquer grains are hardened without deformation of shape by removal of solvent in a series of evaporators. The liquor is removed from the powder, and the powder is washed. The powder is separated in various sizes by a screening operation. Coating stills are used to impregnate the hardened powder with nitroglycerine (increases energy potential) and coat it with deterrent (modifies burning rate). The main function of the dryers is to dry the powder to a uniform moisture and volatile content. Here the powder is also "glazed" with graphite (improves flow characteristics) and salt coated (to adjust ballistics performance). The powder is then blended to improve ballistics uniformity and to meet ballistic requirements of the specific products required.

Inherent Process Safety Advantages

One of the main advantages of the continuous process is increased safety. There are certain design requirements in the continuous process which have resulted in an inherently safe operation. (1) To reduce the hazard created by handling dry nitrocellulose or powder, all operations up to the final dry processing (75% of the handling) are carried out under water. (2) To further reduce handling hazards, all transfers of powder throughout the wet stages are carried out by pumping a water slurry of the material. (3) One of the principal problems in the development of nitrocellulose smokeless powder is that of stabilizing nitrated cotton so as to prevent spontaneous ignition during storage. A superior chemical stability is attained by the addition of a stabilizer to solvated nitrocellulose. Since the molecular dispersion of nitrocellulose in a solvent is an essential step in the grain formation process, it can be seen that (by using solvent containing the stabilizer) stabilization is a natural consequence and that it is not necessary to supply completely purified (stable) nitrocellulose for the BALL POWDER process. The purification and stabilization of nitrocellulose in the BALL POWDER graining operation make it possible to rework and restabilize deteriorated smokeless powder stocks. (4) The potential energy of BALL POWDER is adjusted in the coating operation by the addition of nitroglycerine. The nitroglycerine is added to the propellant in ranges of 10 to 40 percent. An emulsion of nitroglycerine, solvent and water is added and the powder absorbs the nitroglycerine and solvent. Nitroglycerine is a violent explosive and is sensitive to heat and shock. In order to reduce the sensitivity and decrease the hazards of handling nitroglycerine, it is used in solutions with an equal quantity of solvent. A 50% nitroglycerine and 50% ethyl acetate solution when ignited usually burns rather than detonates and cannot be detonated with a blasting cap. The chief danger in the handling of this solution is that ethyl acetate is volatile and upon evaporation will leave a residue of nitroglycerine having its original sensitivity. (5) Drying, which is accomplished quickly in a continuous dryer, entails a minimum amount of powder in process at any time and a minimum exposure of operation personnel at this stage of the process.

Slurry Pump Design Criteria

In the pumping of powder slurries as is performed throughout the wet stages of the process, certain design requirements must be met. It is necessary when pumping slurries of explosives or liquids with a possible explosive containment to avoid circulating explosives behind the impeller or running the pump with inadequate water flow. For this reason open impellers without balance holes are required. Also, a special teflon lantern ring is used. Seal water is introduced at a minimum rate of 1 gpm through a thermal flow switch, along the impeller shaft, and into the casing. The flow switch is interlocked with the motor starter for the pump so that in the event of low seal water flow, the pump will be automatically shut off. The flow switch is calibrated with an in-line rotameter which can also be used to troubleshoot operational problems. The rear seal ring on the pump is removeable so that the entire area behind the impeller may be flushed. Heavy duty thrust bearings are required for all pumps. Pump design is back-pull-out to avoid disconnecting piping when removing the pump impeller.

Equipment Failure Protection

Although the BALL POWDER process is a simple one, a certain

amount of mechanical equipment is necessary.

Mechanical failure in equipment in many cases can create a safety hazard by generating heat which in turn can cause ignition or detonation. By sensing temperature and interlocking the sensor to sound an alarm and shut down the equipment, it is possible to eliminate some hazards associated with mechanical failure. Examples of such interlocks in the grain formation stage of the process are: (1) high temperature shutdowns on lacquer pumps, (2) low flow alarms on mixer journals, and (3) centrifuge bearing alarms.

A continuous mixer is used to dissolve nitrocellulose in ethyl acetate to form lacquer. Lacquer pumps transfer lacquer from the continuous mixer to the grainers. The temperature of the lacquer is already above ambient and any unwanted frictional heat can cause a temperature excursion and possible ignition. The possibility of ignition will be eliminated by placing a temperature probe at the packing gland assembly which will shut down the pump when over-temperature is reached. Similarly, the centrifuge, which separates powder and water, has bearings which on failure could overheat and ignite the powder. A temperature sensor can shut down the centrifuge when the bearing temperature rises. The mixer bearings are handled in a slightly different way. The bearings are cooled by flowing water past them from a constant head water supply tank. If water flow was lost it would be possible for the bearings to overheat, thus creating a very hazardous condition. To eliminate this possibility, a flow switch is installed in the water line to the mixer bearings. When flow is interrupted, an alarm is sounded so that an operator can take corrective action before bearings overheat.

As previously stated, the purpose of the solvent evaporation systems is to remove the solvent from the lacquer particle without deforming the spherical shape of the particle. The loss of the liquid level in the evaporators will result in a very sensitive lacquer and could result in thermal ignition of any residue on the evaporator walls. Level controls are interlocked with the steam supply to the evaporators to prevent thermal ignition from occurring.

In various applications in BALL POWDER facilities, non-explosion-proof vibrating feeders are used to convey powder. At the present time, it is impossible to obtain an explosion-proof feeder. The feeder's electric coil can burn out causing sparking within the enclosure and possibly igniting powder or solvent vapors that have accumulated. To avoid this possibility, air purge systems are installed to prevent the buildup of explosive or flammable material. The air purge systems are also interlocked with the equipment so that in the event of lost air flow, the vibrator will be shut down and an alarm sounded.

Manufacture and Transfer of Nitroglycerine

As defined in the coating operation, nitroglycerine is added to the hardened BALL POWDER to increase its energy potential. Nitration in the nitroglycerine facility is accomplished by drawing glycerine to an injector, the motive fluid for which is nitrating acid. The acid is precooled to approximately 0°C upstream of the injector and serves as both a reactant and a heat sink for the exothermic reaction. Even with the cooling provided by the acid, the nitration temperature is relatively high, 47°C. At this temperature, the reaction is essentially completed in the injector. The reaction temperature is maintained at about 47°C via a pneumatic control valve, which modulates the flow of glycerine to the injector. Insufficient acid flow to the injector will result in a temperature excursion (loss of heat sink) and a loss of vacuum in the nitrator. The temperature excursion may result in pressure rupture of either an acid rotameter, an injector line, or the nitroglycerine cooler due to the rapid evolution of decomposition gases. A vacuum breaker valve is provided to vent the injector vacuum and thus stop the flow of glycerine in the event the vacuum drops too low.

The injector is also equipped with a high and low temperature control loop which will stop the nitration in the event of a temperature excursion. However, the potential for detonation may also exist in the area of the cooler downstream of the injector due to a gradual loss in acid flow resulting in the loss of turbulence in the injector throat. This loss of turbulence will in turn result in an inadequate mixing of reactants and in localized "hot spots." The potential exists in this situation for thermal initiation of coalescing nitroglycerine droplets as either a fume-off or detonation. The temperature sensors located here will alert the operator that the acid flow has decreased before it can be sensed at the injector. After the nitroglycerine is manufactured, it is combined quickly with ethyl acetate to reduce the sensitivity and decrease the hazards of handling nitroglycerine. Therefore, only a small amount of detonable nitroglycerine is in the system at any one time.

The transfer of nitroglycerine presents several potentially hazardous conditions for which design criteria has been established. For example, no screwed connections for fittings are allowed in nitroglycerine service and all welds are inspected for cracks. This reduces the probability of detonation by eliminating pinch points. The handling hazards of nitroglycerine are further reduced by educting, rather than pumping, the solution into the coating still with water. Further, the eduction line is equipped with a flow switch so that in the event that water flow is interrupted and suction is lost, the line will be automatically flushed.

Continuous Propellant Drying

The dry processing of the powder is more hazardous than those described previously for the wet stages of the process. In the continuous drying of BALL POWDER, dewatered powder is fed into a series of vibrating pan conveyors. Fresh air is preheated and blown into an air distribution manifold where tubes direct the air flow into the powder. An aerated bed is obtained by controlling the powder flow rate, the pan conveyor vibration speed, and the heated air supply. The air supplies the heat for evaporating the water. The water vapor is removed with the exhaust air by vacuum transfer.

The air and water temperatures are controlled at 77°C to prevent high temperature which could cause ignition of the powder. The instantaneous flash point of the powder is about 165°C. Only the steam in the heating coils exceeds 160°C. Therefore, it is a standard practice to locate the heating coils and air blowers at a safe distance from the dryer inlet to prevent the possibility of high temperature steam from coming into contact with the powder. As an additional safeguard, fusible link devices installed in the exhaust ducts will automatically activate the closing of the control valve in the steam line feeding the air supply unit before the air temperature becomes excessive. Also, the temperature controls for the supply units are interlocked to prevent the supply blower motors from starting if there is steam in the heating coils. This prevents an initial high rate of rise of temperature.

A normal problem encountered in the manufacture of BALL POWDER is the generation of a certain quantity of "fines" (less than 0.009" diameter). In order to supply enough heat to maintain the established drying temperature of 77°C, a certain quantity of "fines" and low density powder may become entrained in the exhaust air. A serious hazard will exist if the entrained powder falls out of the exhaust air stream and accumulates in the ducts or the exhaust fan shroud. To prevent this problem, the exhaust manifold of each dryer unit has a large cross sectional area at the exit port. This reduces the exit velocity of the exhaust air, minimizing entrainment of solids. As the exhaust air leaves the exit port, the cross sectional area of the duct decreases, which increases the air velocity above the minimum transport velocity. This prevents solids from falling out of the air stream and settling in the duct. The air is then pulled into high efficiency cyclones which separate any entrained powder from the air. In order to insure high separation efficiency, fresh bleed air is added to the exhaust air to maintain a constant air flow (the pressure drop across the cyclone varies as the square of volume handled). The air exhausted from the cyclones feeds into a single wet scrubber which scrubs the exhaust air with water to remove any graphite or traces of powder not removed by the cyclone. This type of system allows high volumes of drying air to be safely and efficiently used for both high density and low density powder. (Low density powders are more easily entrained in the exhaust air.)

The fresh bleed air introduced into the dryer exhaust ducts is heated to prevent the condensation of any nitroglycerine vapors in the exhaust air (a small amount of nitroglycerine evaporation from the powder in the latter dryer zones is not unusual). The heated air assures that any nitroglycerine in the exhaust air remains in the vapor state until it can be removed by the wet scrubber.

In areas where dry powder is handled, the buildup of static electricity presents a safety hazard. A static charge could possibly ignite or detonate propellant in the dry state. For this reason, materials used in these areas are conductive and properly grounded. Flooring and footwear must also be conductive as outlined in the Department of Defense Safety Manual. Conductivity tests are made periodically to insure that there is proper protection against static charge buildup.

Nitrocellulose smokeless powder is an extremely electronegative material, and therefore, produces static electricity when brought into contact and separated from almost any other material. During the course of drying, and handling the powder after drying, BALL POWDER could acquire a considerable charge. To prevent this from occurring, BALL POWDER is "glazed" with graphite and handled in grounded containers so that it cannot accumulate an appreciable static charge.

Fire Protection

The fire water distribution system is kept pressurized by means of a jockey pump. In the event of a sprinkler trip, three large diesel driven pumps come on line as needed to supply the tremendous quantity of water which may be required. In order to prevent the water pressure from dropping, which would result in low water flow to the sprinkler nozzles immediately after they are opened, an electric switch panel allows the diesels to be started directly from an electric notifier system. This permits the diesels to get the start signal about the same time water begins to flow which reduces the unwanted initial pressure drop.

In areas where the possibility of fire is the greatest, a high speed ultraviolet detection system is employed. The ultraviolet "eyes" detect the presence of a flame and are interconnected with a high speed water deluge system. In some areas the fire detection devices have process shutdown capabilities. The detectors are situated such that the entire hazardous area can be monitored at all times. They are equipped with air shields that not only cool the detector but also blow clean air across the face of the lens, keeping any dust from accumulating on the lens, which assures detection capability. Further, the detectors have a built-in test lamp which serves as a check of the detector tube module and also makes it possible to check for lens obscuration. The system is one that is highly reliable and which will sound an alarm in case of a broken wire or malfunction of any critical component of the system.

The fire protection system consists not only of a deluge system activated by the ultraviolet sensors, but also of fusible link type fire systems. The deluge system has a trip mechanism from mercury checks activated by heat-activated-devices, a manual release on the deluge valve, a pneumatic remote trip station, and an electrical push button along with the electrical trip mechanism from the U/V detectors. The remote trip stations are located by escape routes so it is possible for the operator to trip the systems as he exits the building without exposing himself to further danger.

Also used to help contain potential fire is fireproofing of steel tanks. Any steel tank containing flammable liquids has the steel supporting members covered with fireproofing material having an underwriter's laboratory rating of at least two hours. This will prevent a tank from collapsing and fueling a fire.

Any areas which have flammable liquids are equipped with a foam fire fighting system. The foam system is activated by fusible links or manually. The fire truck has a foam eductor system which can be used to extinguish flammable liquid, or Class II fires. The foam is educted out of five gallon buckets and mixed with the water stream in the eductor.

Summary

Prevention of loss of personnel or property is more than just a social or moral obligation. Safety is as critical to the efficient operation of a chemical plant as any other function. The manufacture of explosives involves special hazards which must be minimized through intelligent safety engineering.

Received November 22, 1978.

9

Explosion Suppression of Large Turbulent Areas

WILLIAM A. CROSLEY

Detector Electronics Corp., 7351 Washington Avenue South, Minneapolis, MN 55435

Explosions have been occurring since the world was formed. In the last century explosions of man-made origins have been more common and most costly. To many people it is a disaster such as an earthquake or a tornado, and the only protection is to hope it doesn't happen - and feeling there is no way to stop the disaster once it starts.

There is enough knowledge and equipment available today to react and stop certain explosions even after they have begun. An explosion that is preceeded by an ignition is a candidate for an explosion that can be suppressed before major harm is done.

Explosion suppression systems are and have been used in all parts of the world for some time, but these systems are usually protecting very small and unobstructed areas, where a mechanical pressure sensor can be used successfully to detect the initial pressure from an explosion.

Large-scale suppression tests in obstructed areas had never been tested until the U.S. Coast Guard conducted tests in the tanker S.S. Texaco at Mobile Bay, Alabama utilizing ultraviolet detection to actuate the suppression systems.

So that you can better visualize the types of explosions we were attempting to suppress, we have a movie film of several unsuppressed explosions.

These are slow motion pictures taken through the port hole at a camera speed of 64 frames per second, looking at the tanker's pump room where 12 pounds of liquid propane were allowed to vaporize and mix with the air to form a stoichiometric mixture. The fuel was then ignited by means of an electric arc or a hot wire.

Note the blue flame front indicating the air fuel mixture was truly a stoichiometric one. The flame front reaching the obstructions moves faster around the obstructions than the unobstructed portions of the flame front. This visually displays the fact that the flame front of an explosion moves faster when obstructed than in an empty volume. We will see a slide after the movie showing the over-pressure developed faster in this obstructed area opposed to the pressures developed in a considerably less obstructed area.

0-8412-0481-0/79/47-096-179$05.00/0

The size of the compartment was approximately 32 feet wide, 16 feet long and 40 feet high. The total volume was approximately 18,000 cubic feet. The over-pressure of the unsuppressed explosions was 12 psi.

These unsuppressed tests were conducted to develop data such as over pressures developed when a stoichiometric mixture of propane is ignited as well as how fast the flame front advanced in an obstructed area.

These next scenes are not sequences of an explosion, but of a burning torch being thrown into an open cublicle with an open pan of gasoline on the floor. In the cubicle is a UV fire detector and a 10-pound bottle of Halon 1301, an extinguishing agent.

The purpose of this scene is to better acquaint you with the speed of the system visually, so that you can better understand the rest of this speech today.

The UV detector is in the upper left corner of the cubicle, viewing only the area inside the cubicle. The 10-pound bottle of Halon is in the opposite corner. When the flames from the torch enter the 90° viewing cone of vision of the detector inside the cubicle, the detector instantly signals a relay in its controller which closes and causes a detonator cap to rupture the disc on the Halon bottle - releasing the Halon. The Halon suppresses the flames on the torch and the unlighted torch falls harmlessly into the pan of gasoline.

The purpose of the explosion-suppression tests at Mobile Bay was to determine (1) if it was possible to suppress an explosion in such a large volume, (2) to determine the most suitable type of detection and (3) to evaluate various types of extinguishing agents.

The tests proved an explosion could be suppressed if detected in its early stages with extinguishing agents in sufficient quantity, reaching the flame front in less than 100 milliseconds after ignition.

Pressure sensors were used initially but were not successful because of their slow response and the rapid pressure buildup due to the speed of the flame propagation. Ultraviolet (UV) detectors were determined to be the only type suitable for this purpose.

Of the extinguishing agents, Halon 1211, 2402 and 1301 were successful, but in fairly high concentrations. Water was partially successful and dry chemical was the most successful on a concentration basis.

I will now describe how the tests were conducted and the information obtained from these tests. First, let's discuss the detector portion of the system.

The following considerations are of utmost importance for the detection portion of the system:

1. In a large obstructed area where there will be turbulence, the only type sensor available is a radiation type sensor because of the increased speed of the flame propagation. It was proven during these tests that pressure sensors

1. Continued.
   are not suitable.
2. Once the detector sights the fireball, it must respond in milliseconds.
3. To suppress an explosion, the detectors must be positioned so that a fireball occurring anywhere within the protected space will be detected by at least one detector in less than 75 milliseconds after ignition. If it takes longer, the explosion will be fully developed before the extingusihing agent can reach it, and the result will be a partial suppression, or very likely, no suppression.
4. The detection system must be protected from false or spurious signals.
5. The detection system must incorporate safeguards against system failure and provide an early warning system to indicate if the detector is becoming less sensitive, due to an electronic malfunction or contamination of the viewing window.

The detector best suited to meet these conditions is a radiation type sensor. Ultraviolet (UV), infrared and visible radiation are generated when combustion produces a flame and all three types of radiation sensors respond to the radiation from the flame.

The visible and infrared sensors are fast, but are subject to many false signals from artificial light, sunlight, hot bodies and other heat producing bodies. The ultraviolet (UV) sensor is fast in response and is affected by few extraneous signals which are controllable. The UV sensor must be the type that only responds to a narrow band of UV, from 1850 Angstroms to 2450 Angstroms. This is considerably below the UV radiation wavelengths from the sun and artificial lights.

A basic UV detection system consists of these three basic components:

A UV detector having a 90° cone of vision. The UV generated in that viewing area will cause the UV detector to send a voltage pulse to the signaling process section of the electronic amplifier/controller. Once UV from the exploding fireball enters the sensor's 90° cone of vision, the solid-state switch will close in less than 10 milliseconds, which will actuate the extinguishing system.

The UV detection system like any man-made product, is subject to failure. These failures could be detector tube failures, electronic or wiring failures, or optical system failure or contamination.

There are UV fire detection systems available that have the capability of self-examination for failures. This supervisory circuit is referred to as Automatic Optical Integrity.

Let's examine a basic UV system that incorporates this Automatic self-examination feature. UV from an exploding fireball will enter the optical surface of the UV sensor and generate a voltage signal to cause the controller's relay to actuate and release the extinguishing agent to suppress the explosion. Also, a UV signal generator, in this case a very small UV source lamp, screened from the sensor will pulse briefly every few seconds. This small level of UV radiation can only reach the sensor by passing out through the viewing window and reflecting off a beveled ring and then re-entering the detector housing in the same location where the UV from a fireball would enter. The momentary reaction of the sensor from the UV source lamp is a signal to the system that everything is in operation from the detector window through the UV sensor and through the electronics. If the UV from the source lamp fails to pass through or the sensor fails to respond, a fault relay de-energizes, warning the operator of a malfunction which will normally be a contaminated viewing window. Once the window is wiped clean the fault relay can be cleared indicating the system is no longer sub-marginal. This type of self-examination is known as Automatic Optical Integrity.

This is a cut-away view of the detector inside its explosion-proof housing. The UV source lamp generates UV which is isolated by this barrier from the detector tube. The UV cannot pass through this barrier but must pass through the quartz viewing window, and at that point strike the UV detector sensor. If the quartz window eventually becomes contaminated, all of the UV from this source will not be able to penetrate the surface of the quartz window, thus reducing the signal to the detector, causing the system to go into a fault alarm condition.

Now I will describe the U.S. Coast Guard test site and the explosion suppression systems used. The pump room measured 40 feet from the bilge level to the top hatch, and 32 feet from port side to starboard, and 16 feet from fore to aft. The total volume, including two first level wings, minus the space consumed by the pumps and other obstructions, was 18,000 cubic feet.

Two explosion suppression systems were tested. They were based on the following principles:

1. UV detector sees incipient explosion.
2. It signals the control circuitry which fires an electrically actuated blasting device.
3. The blasting device ruptures a restraining diaphragm, causing the release of suppression agent which is under 600 psi dry nitrogen pressure.
4. The driving force of the nitrogen forces the agent out of the container and propels it toward the flame front.
5. The flame is extinguished; the explosion suppressed.

One system consisted of cylindrical cannons to contain the agent. The driving force was provided by pressurizing the charged cannon to 600 psi with dry nitrogen. Again, UV detectors were used to sense the fireball, and the suppression agent used was the dry chemical known as Purple K.

The other system consisted of spherical high-rate discharge extinguishers to contain the agent. The driving force was provided by pressurizing the charged extinguisher to 325 psi with dry nitrogen. The UV sensor was used for the detection of the fireball, and various suppression agents were used in these extinguishers. They included water, Halon 2402, Halon 1211 and Halon 1301.

Nine UV detectors were installed throughout the entire pump room, the same as would be necessary in an actual installation. If any one of these detectors sees UV radiation, it will fire all of the cannons. The circuitry was designed to indicate at any time if there was an electrical malfunction in the wiring to any of the cannon detonators.

The extinguishing cannons or spheres were located throughout the entire pump room above and below the deck plates as were the nine UV detectors. In this manner a growing fireball could be detected at any point within the entire pump room within milliseconds after ignition.

Two cannons were located in the top hatch propelling dry chemical down.

At the third deck level there were two cannons on the port side and two on the starboard bulkheads.

At the second level there were four additional cannons with the nozzles at a 45° angle firing directly over the pumps in the pump room.

There were several cannons on the bottom level firing below the deck and just above the bilge water.

The pump room was obstructed with the normal equipment, pumps, valves, pipes and ladders, the same as when it was in service.

If the entire pump room were filled with a stoichiometric mixture of propane and air throughout the entire volume, unsuppressed explosions would produce pressure of approximately 120 psi, enough pressure to blow the ship out of the bay. Naturally, we could not run tests that could develop these pressures.

Based on the design limitations of the pump room bulkheads, it was decided that the bulkheads could easily withstand 15 psi. Calculations show that if the stoichiometric mixture of a hydrocarbon fuel were placed in 10% of the volume and ignited, expansion of the explosion into the remaining 90% of the volume would limit the theoretical maximum pressure to 12 psi, and not affect the bulkheads. With this reasoning 12 pounds of propane were used.

Also, a successful suppression should hold the maximum explosion pressure to less than 1 psi. Thus, it was reasoned that the total volume would produce realistic results for explosion suppression purposes. The actual successful suppressions that followed were all less than 1 psi and as low as .2 psi.

The first tests used pressure sensors for the detection system, but they failed to respond fast enough to suppress the explosion. It was suspected that the obstructions in the pump room increased the speed of the flame propagation beyond the capability of a pressure sensor.

To prove this, the bilge water was allowed to rise above the top of the pumps and other major obstructions. Additional unsuppressed tests were conducted and the pressure rise was much slower as suspected in an unobstructed area. This proved that pressure sensors can be used in empty spaces but not in obstructed large spaces. At this point the pressure sensors were no longer used and the UV fire detection systems were now used.

During the explosion suppression tests, the UV detector responded between 18 and 25 milliseconds after ignition. The first contact of the extinguishing agent with a fireball occurred in less than 100 milliseconds after ignition, and the peak of the suppressed curve occurred within 150 milliseconds at an over-pressure as low as .2 psi.

In analyzing these tests, the following factors were considered: Size of fireball at agent contact, degree of agent breakdown, amount of burning after the initial suppression, and maximum pressure developed. The analysis indicates that successful suppression requires the listed minimum application densities for the suppression of explosion of a propane/air mixture in obstructed spaces similar to a ship's pump room.

In summary, the amount of agent required to successfully suppress the explosion in the pump room or similar 18,000 cubic foot obstructed area, is as follows:

Water required a concentration of more than .15 pounds per cubic foot.

Halon 2402 required more than .12 pounds per cubic foot.

Halon 1211 required between .06 and .09 pounds per cubic foot.

Halon 1301 required between .05 and .08 pounds per cubic foot.

Purple K (dry chemical) required more than .007 pounds per cubic foot.

This illustrates that the dry chemical was more than 10 times as effective as the most efficient Halon.

Water was the least efficient and was the only test that was not completely successful. There were not enough containers to be able to reach a .15 pounds per cubic foot density, but there was a partial suppression at .066 pounds per cubic foot density. The .15 pounds per cubic foot was determined by extrapolation.

When the Halon density was not sufficient to suppress the explosion, there was agent breakdown evidenced by the observation of orange-yellow smoke and severe afterburning.

In conclusion, these tests indicate that it is possible to suppress certain explosions in large obstructed areas.

Pressure sensors are not suitable as explosion suppression systems in large obstructed areas, but the basic ultraviolet (UV) detection system available today as a fire detector is capable of being used as an explosion detector, providing the exploding fireball is in the cone of vision of one of the detectors within the first 75 milliseconds. The extinguishing agent used must make contact with the fireball within 100 milliseconds after ignition. The density of the extinguishing agent required will vary with agent used. In these particular tests, dry chemical was far superior on a weight per unit volume basis by a magnitude of 10 times or greater.

The movie film that now follows shows several unsuppressed tests where data was recorded and then a suppression by each of the 5 agents used for suppression.

Abstract

Explosion suppression for large volumes in which turbulence can be expected is one of the greatest challenges in the fire protection field today, and the need for such a system is great. There are explosion suppression systems in service throughout the world, but in almost every case they are in very small volumes of unobstructed areas, where the mechanical pressure sensor can be used successfully to detect the explosion. Large-scale suppression tests in obstructed areas have never been tested until recent U.S. Coast Guard tests were conducted on an ocean oil tanker utilizing ultraviolet detection systems to actuate dry chemical, Halon and water suppression systems. The purpose of these tests was to determine the most suitable type of detection, and to evaluate various types of extinguishing agent. Pressure sensors were tried initially, but were not successful because of their slow response, and ultraviolet (UV) dectors were finally determined to be the only type that were suitable for this purpose. Of the extinguishing agents, Halon 1211, 2402 and 1301 were successful, but in fairly high concentrations. Water was partially successful and dry chemical was the most successful on a concentration basis. Emphasis should be made that this was a research program to gain additional information for the design of a complete suppression system with a capability of suppressing explosions in large areas as well as small enclosures. Work is continuing in our company to develop these systems. The ultraviolet explosion detectors and the suppression systems are commercially available but must be engineered and designed for each hazardous application.

RECEIVED November 22, 1978.

10

Rapid Suppression of Explosive and Incendiary Fires

CHRIS C. ELKINS

Day & Zimmermann, Inc., Operating Contractor of Lone Star Army Ammunition Plant, Texarkana, TX 75501

In 1951 Day & Zimmermann, Inc. was awarded an initial contract to operate the Lone Star Army Ammunition Plant. Fire protection systems for equipment and processes consisted of fuzeable links and quartzoid bulbs for sensing fires. In the early 1950's, a plan was initiated to replace the existing systems with the improved Heat Actuating Device. The H-A-D greatly improved the effectiveness of the deluge systems and was considered adequate at the time.

Introduction of new materials and processes created new problems due to the frequent fires and explosions. These problems surfaced in the late 1960's and early 1970's. A search for an ultra-high speed fire protection system was begun during this period. Ultra-high speed is a term that measures in milliseconds as compared to high speed, such as the H-A-D which measures in seconds.

It has been said that necessity is the Mother of Invention, but at Lone Star Army Ammunition Plant, it was the Mother of Motivation. The cost of repair of fire damage was running high and our people had become very skittish. The feeling of urgency compelled us to find or design a better fire protection system.

In 1972, Day & Zimmermann engineers started a test program. One of the first steps was to select the type of sensor or detector. Consideration was given to infra-red, ultraviolet, and audio signals. After careful study of the alternatives, the ultraviolet was chosen, primarily because of available components and that it would not be affected by sunlight. Audio was discarded because of frequent thunderstorms during the rainy season.

The second major component selected was the deluge valve. It was to be a fast-operating valve located as close to the nozzle as practical. Automatic Sprinkler had such a valve that was built into the nozzle. This would provide a wet system in close proximity of the fire initiation point. It would be pilot-operated for fast response.

0-8412-0481-0/79/47-096-187$05.00/0

The third major component would be the controller that takes the signal from the ultraviolet sensor and amplifies it to a power voltage that operates the valve. The controller was constructed of standard electrical components that were on hand. In addition to opening the valve, it would shut off the power to the machines to prevent additional damage and would also sound an alarm.

These components were assembled, and a test was conducted to compare the new system with the H-A-D and quartzoid systems. The first series of tests used various test materials such as TNT, propellant powders, delay compositions, and black powder.

Electric matches were tied into a common power source to give simultaneous ignition of the test samples. Movie film was used to record the test results. Frames of the slow motion film were counted to determine the approximate response time of the systems.

The results of the first series of tests indicated that the ultraviolet sensor deluge system was considerably faster than the H-A-D and the older sensor, the quartzoid. In fact, the quartzoid sensor never actuated during these tests. The response time from initiation of the test sample until deluge water was on the fire ranged from 300 to 400 milliseconds for the ultraviolet system. This response time was adequate to suppress fires from most of our chemicals and explosives. The one exception discovered during the tests was a black powder fire. Unlike all the other materials, the black powder sample was completely consumed by the fire. Even though the system failed for black powder, the series of tests gave merit to the ultraviolet sensor deluge and provided additional valuable information which approximated the response time of the controls and valve.

A second series of tests was conducted specifically to improve the response time of the system and extinguish the black powder fire. The only change in the test set-up was the use of solid state controls in lieu of the conventional ones used during the first tests. The retests were made and the response time was reduced in half; but again the entire sample was consumed and the system was not considered to be effective.

Before the third series of tests was made, additional research was required to find or develop an ultra-fast opening valve. This was the only major component that could be replaced to further reduce the response time. Information from Detector Electronics Corporation introduced us to a valve manufactured by Grinnell, called the Primac valve, which seemed to fit the bill. The valve uses a primer detonating device with redundant detonators to blow the valve open. The same electrical signal initiated by the ultraviolet sensor could now be used to actuate the detonators, thus further reducing lost motion.

The test set-up was modified slightly from the previous test in that the black powder was spread along a path one inch wide and 30 inches long with the electric match at one end and a cup with several ounces of powder at the other. The intention was to measure the distance the fire burned before or if the fire was extinguished. The tests were recorded on Fastex film taken at 4200 frames per second. The reponse time from ignition until water was on the fire was 62 milliseconds.

The response time had again been cut in half. Nineteen inches of black powder were consumed in the test. In order to get the water on the fire a little faster, the line pressure (approximately 75 psi) was increased to 100 psi and the system was retested. This time, only six inches of powder were consumed. This system is considered highly effective and is in use today at our black powder loading operation.

Day & Zimmermann has installed approximately 90 ultraviolet sensor deluge systems in the Lone Star and Kansas Army Ammunition Plants. The Primac valve is only used in the black powder operation because of its higher cost. All other materials use the system as described in the second series of tests and is considered highly effective.

Because of the nature of our business, we continue to have the fires but the cost is generally negligible in comparison to the cost of fire damage when the old systems were in use. Production downtime has been greatly reduced, and the operators working the lines feel much safer.

Certainly the ultimate fire protection system has not been achieved, but we feel at this time that we are pushing the state-of-the-art.

RECEIVED November 22, 1978.

11

Laboratory Design and Operation Procedures for Chemical Carcinogen Use

MANUEL S. BARBEITO

National Cancer Institute, Office of Research Safety, Bldg. 13, Room 2E47, 9000 Rockville Pike, Bethesda, MD 20014

Interim safety standards (1) were developed by the National Cancer Institute (NCI) for the intramural laboratories for research work involving chemical carcinogens regulated by the Department of Labor (2). The standards are also used as guidelines for suspected chemical carcinogens and other toxic substances. The interim standards were issued to comply with the policy of the NCI that all of its research programs be planned and implemented to prevent exposure of personnel to hazardous material, minimize environmental contamination and provide product protection when required. The NCI standards are based on the proposed Department of Health, Education, and Welfare Laboratory Chemical Carcinogen Safety Standards, prepared by the DHEW Committee to Coordinate Toxicology and Related Programs. When DHEW issues the standards, they will supersede these interim NCI standards. A summary of these interim standards for research involving chemical carcinogens follows:

Medical Surveillance Program

A preassignment physical examination is provided each person planning to work with chemical carcinogens. No employee is required to participate in the Medical Surveillance Program; it is strictly voluntary. This examination establishes a baseline against which changes can be measured or to determine if there exists any medical or other condition that could compromise the employee's health in the work situation. A periodic examination based on age is also provided for employees using chemical carcinogens.

Frequency of Medical Examination for Employees Working with Chemical Carcinogens

AGE (yrs)	FREQUENCY (yrs)
20 - 30	5
31 - 40	2
over 40	1

These examinations seek to determine any changes in the medical status of employees as the result of his/her work.

Personnel Practices

Protective Clothing. Protective clothing for laboratory personnel is provided daily. As a minimum, it consists of a fully fastened, long laboratory coat. The protective clothing (laboratory coat, jumpsuit or pants/shirt, smock, and gown) is not to be worn outside of the work area, Figure 1.

If clothing is contaminated with chemical carcinogens, it is not sent to commercial or in-house laundry facilities until decontaminated, if feasible, or it is discarded in a safe manner. Disposable laboratory garments are often preferred for use in laboratories engaged in research using chemical carcinogens. Gloves are worn that meet the specific research needs and are resistant to solvents (3) and impermeable to the chemical carcinogens in use. It is recommended that animal handlers or others entering areas where animals are treated with carcinogens or fed diets containing chemical carcinogens be provided daily with a complete clothing change, including shoes, boots, head cover, and gloves.

Personnel protective equipment may be used in certain circumstances where exposure to airborne particulates contaminated with chemical carcinogens could occur. In those situations, personnel should be equipped with a complete clothing change, as well as respiratory protection selected on the basis of work performed, type of chemical used, and containment equipment. The respiratory protection may be a face mask, respirator [selected from those approved by the National Institute for Occupational Safety and Health (NIOSH)] (4, 5), or emergency breathing air system. In the latter case, a head hood or a complete protective suit may be used with a breathing air supply system, Figure 2. When respiratory protection is used, it should be decontaminated daily. However, the use of respiratory protection as the primary means of preventing exposure of laboratory personnel should be avoided whenever possible.

Research work should be conducted in primary containment safety equipment, such as Class I or Class II (Type B) biological safety cabinets, a chemical fume hood, a gastight glove box, or other specially designed equipment. Research work with chemical carcinogens on laboratory bench tops with the use of protective clothing, including respirators, is not a good alternative to conducting all work in primary enclosures. The major problems arise from contamination of the entire laboratory and from exposure of personnel when disrobing following the use of protective equipment.

Showers. When personnel may be exposed to airborne particulates contaminated with chemical carcinogens, they should shower after exit from the work area. Also, all personnel should shower immediately after an overt exposure to a chemical carcinogen.

Pipetting. All liquid transfer operations are performed with mechanical pipetting aids, Figure 3; also see the "Source" table (6). Oral pipetting is not performed in a laboratory engaged in research work with chemical carcinogens, suspect carcinogens, or other toxic substances (7).

Eating, Drinking, Smoking. There should be no eating, drinking, smoking, chewing of gum or tobacco, application of cosmetics, or storage of food in areas where chemical carcinogens are used.

Personal Hygiene Practices. All personnel should wear disposable or other type gloves resistant to research materials (3) and wash their hands immediately after removing gloves following the completion of any procedure in which chemical carcinogens are used. A hand washing facility should be made available within the laboratory.

Operational Practices

Access Control. A sign should be posted at entrances to all work areas where carcinogens are present (2). The National Cancer Institute uses a sign: "DANGER, CHEMICAL CARCINOGEN, AUTHORIZED PERSONNEL ONLY." Generally, access is authorized by the laboratory supervisor. In addition, entrance procedures and clothing requirements should be permanently displayed at the main access point. To prevent any untoward incident, all visitors should be escorted in laboratories engaged in research work involving known or suspect chemical carcinogens or other toxic substances.

Work Surfaces. All horizontal work surfaces (bench tops, containment cabinets or fume hoods) should be protected with impervious material to prevent contamination of the work surfaces with chemical carcinogens. One of the systems that has proved useful is to use the dry, absorbent polyethylene-backed paper (Benchkote, VWR Scientific Co., Cat. No. 52855, Continuous Sheet Type, Scientific Products Co., Cat. No. P1180). Following contamination, or upon completion of an experiment, or at the end of the day, this protective cover can be rolled up, packaged for safe removal, and disposed of in an appropriate manner.

Primary Containment Cabinets. Primary containment cabinets should be appropriately labeled with the warning: "DANGER—CHEMICAL CARCINOGEN." Procedures requiring the use of containment cabinets are: 1) when using volatile or suspected chemical carcinogens; 2) when procedures result in the formation of aerosols, such as from opening closed vessels, transferring operations, weighing, preparing of feed mixtures, injecting and intubating experimental animals with chemical carcinogens.

Figure 1. Types of laboratory clothing. (Top left) *simple cap,* (top right) *bouffant cap,* (bottom) *hooded cap.*

Figure 1. Types of laboratory clothing. (Left to right, top to bottom) *Fully buttoned laboratory coat, wraparound smock, solid front gown, one-piece laboratory suit, two-piece laboratory suit, heavy duty coverall, simple cap, bouffant cap, and hooded cap.*

Figure 2. Respiratory protective equipment. Full-mask facepiece, half-mask facepiece, air-supplied head hood.
Mine Safety Appliances Company

Figure 2. Respiratory protective equipment. One-piece positive pressure ventilated suits.
Mine Safety Appliances Company

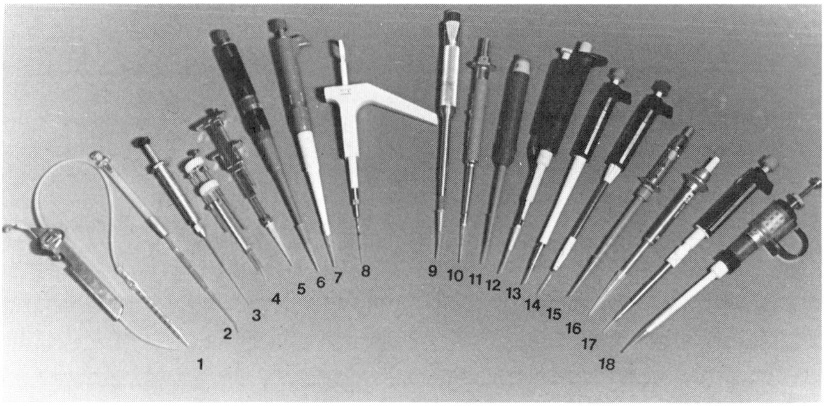

1 Clinac Safety Pipettor
2 Sargent-Welch Pipette Syringe
3 Drummond Dialamatic Micro Dispenser
4 Hamilton Ultra Micro Syringe Pipet
5 Unimetrics Micropipettor
6 Finnpipette (Finnish manufactured)
7 Helena Quickpette
8 Clay Adams-Micro Selectapette
9 Lancer Precision Pipettor
10 MLA Pipettor
11 Eppendorf Microliter Pipet (West German manufactured)
12 Pipetman Ultra Micro Pipette (French manufactured)
13 Oxford Sampler Pipette (8000 Series)
14 Oxford Sampler Ultramicropipette
15 Elkay "Socorex" Micro Pipette (Swiss manufactured)
16 Centaur Pipet
17 Oxford Sampler Pipette, Model Q
18 Gilson/Rainin Pipetman, Digital Dispensing Pipette (French manufactured)

Ultramicro Pipetting Aids

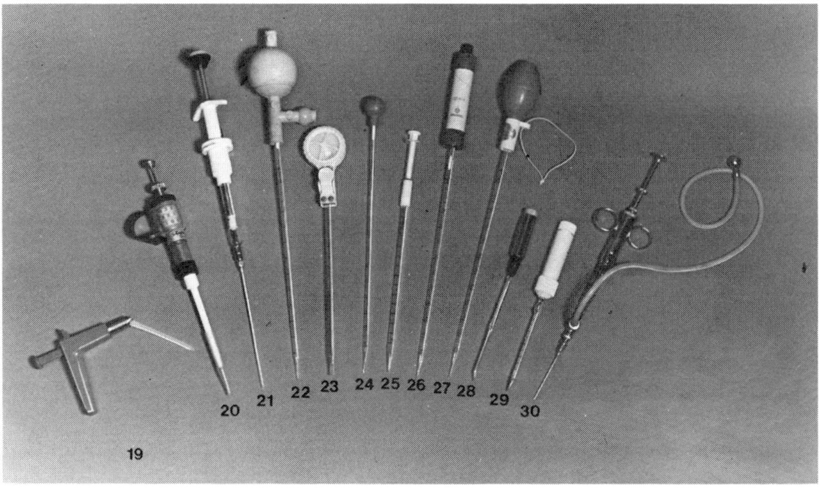

19 Biopipette Automatic Pipette
20 Gilson/Rainin Pipetman, Digital Dispensing Pipette (French manufactured)
21 Manostat Vari-Pet
22 Spectroline Pipette Filler
23 Pumpett 18 (Swedish manufactured)
24 Curtin Matheson Rubber Pipet Bulb
25 Demuth Safety Pipet
26 Nalgene Pipetting Aid
27 Analytic Products Safety Bulb
28 Clay Adams Pipet Suction Apparatus
29 Manostat Accropet Filler
30 Cornwall Continuous Pipetting Outfit

Micro Pipetting Aids

Figure 3. Pipetting aids

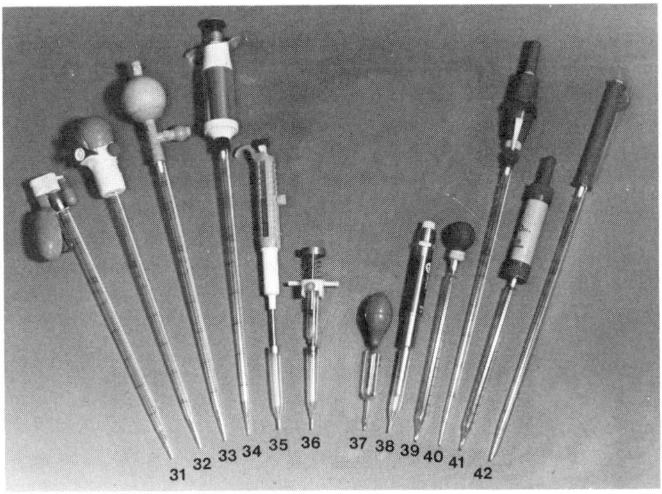

31 National Instrument Micropipet Pipettor (French manufactured)
32 Pumpett 25, Original (Swedish manufactured)
33 Spectroline Pipette Filler
34 Manostat Accropet Filler (Macro)
35 Oxford Macro Set Transfer Pipetting System
36 Lab Industries Repipet Jr Sampler
37 Labindustries Micro/Mac Pipet
38 Centaur
39 Interex Rubber Transfer Bulb
40 Volac Universal Pipette Controller (British manufactured)
41 Nalgene Pipetting Aid
42 Pi Pump (West German manufactured)

Macro Pipetting Aids

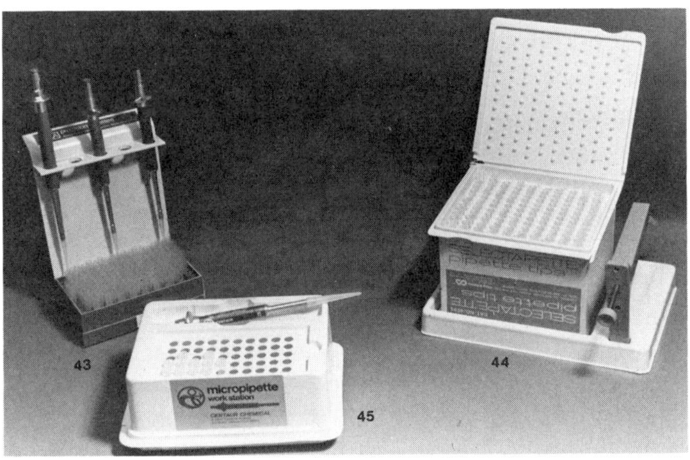

43 MLA Inc. Pipetting System
44 Clay Adams Selectapette Pipet System
45 Centaur Pipet System

Pipetting Aid Systems

46 Drummond Pipet-Aid

Pipetting Aid System

Figure 3. Pipetting aids

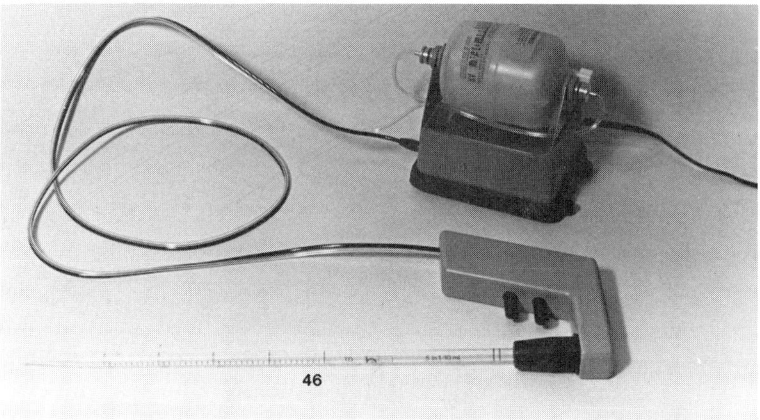

Figure 3. Pipetting aids

SOURCES OF PIPETTING AIDS

Item / Source	Figure Legend Numbers	Price Range*
Analytic Products Safety Bulb Analytical Products, Inc. P.O. Box 845 Belmont, CA 94002	27	C
Biopipette Automatic Pipette Schwartz/Mann Division of Becton, Dickinson & Co. Orangeburg, NY 10962	19	I
Centaur Pipet Centaur Chemical Company 180 Harvard Avenue Stanford, CT 06902	16, 38	G, H
Centaur Pipet System	45	H
Clay Adams-Micro Selectapette Clay Adams Division of Becton, Dickinson & Co. Parsippany, NJ 07054	8	I
Clay Adams Pipet Suction Apparatus	28	C
Clay Adams Selectapette Pipet System	44	I
Clinac Safety Pipettor LaPine Scientific Company 6001 South Knox Avenue Chicago, IL 60629	1	E
Cornwall Continuous Pipetting Outfit Curtin Matheson Scientific Co. P.O. Box 1546 Houston, Texas 77001	30	E
Drummond Pipet-Aid	46	J
Elkay "Socorex" Micro Pipette (Swiss manufactured) Elkay Products, Inc. 95 Grand Street Worcester, MA 01610	15	H
Eppendorf Microliter Pipet (West German manufactured) Brinkman Instruments, Inc. Cautrague Road Westbury, NY 11590	11	H
Finnpipette (Finnish manufactured) Variable Volumetrics, Inc. 17 Cummings Park Woburn, MA 01801	6	I
Gilson/Rainin Pipetman, Digital Dispensing Pipette (French manufactured) Rainin Instrument Company, Inc. 94 Lincoln Street Brighton, MA 02135	18, 20	I
Hamilton Ultra Micro Syringe Pipet Hamilton Company P.O. Box 7500 Reno, NV 89502	4	H
Helena Quickpette Helena Laboratories P.O. Box 752 Beaumont, TX 77704	7	H

Figure 3, cont'd. Sources of pipetting aids

SOURCES OF PIPETTING AIDS (Continued)

Item / Source	Figure Legend Numbers	Price Range*
Curtin Matheson Rubber Pipet Bulb Curtin Matheson Scientific Co. P.O. Box 1546 Houston, Texas 77001	24	A
Demuth Safety Pipet Demuth Glass Division Brockway Glass Company, Inc. Route 1, Box 13 Parkersburg, WV 26101	25	A
Drummond Dialamatic Micro Dispenser Drummond Scientific Company 500 Parkway Broomall, PA 19008	3	H
Manostat Accropet Filler Manostat Corporation 519 Eighth Avenue New York, NY 10018	29, 34	C
Manostat Vari-Pet	21	G
MLA Pipettor & Pipettor System Medical Laboratory Automation, Inc. 520 Nuber Avenue Mount Vernon, NY 10550	10, 43	H, I
Nalgene Pipetting Aid (West German manufactured) Nalgene Labware Division Nalge Sybron Corporation P.O. Box 365 Rochester, NY 14602	26, 41	D
National Instrument Micropipet Pipettor (French manufactured) National Instrument Co., Inc. 4119 Fordleigh Road Baltimore, MD 21215	31	D
Oxford Macro Set Transfer Pipetting System Oxford Laboratories 1149 Chess Drive Foster City, CA 94404	35	I
Oxford Sampler Pipette (8000 Series)	13	H
Oxford Sampler Pipette, Model Q	17	H
Oxford Sampler Ultramicropipette	14	H
Interex Rubber Transfer Bulb Interex Corporation 3 Strathmore Road Natick, MA 01760	39	A
Labindustries Micro/Mac Pipet Labindustries 1802 Second Street Berkeley, CA 94710	37	B (glass) A (rubber)
Labindustries Repipet Sampler	36	E
Lancer Precision Pipettor Sherwood Medical Industries 1831 Olive Street St. Louis, MO 63103	9	H
Pi Pump (West German manufactured) Bel Art Products Pequannock, NJ 07440	42	D
Pipetman Ultra Micro Pipette (French manufactured) Analtech, Inc. 75 Blue Hen Drive Newark, DE 19711	12	I
Pumpett 18 (Swedish manufactured) LaPine Scientific Company 6001 South Knox Avenue Chicago, IL 60629	23	C
Pumpett 25, Original (Swedish manufactured)	32	C
Sargent-Welch Pipette Syringe Sargent-Welch Scientific Co. 7300 North Linder Avenue Skokie, IL 60076	2	D
Spectroline Pipette Filler Arthur H. Thomas Company Vine Street at 3rd P.O. Box 779 Philadelphia, PA 19105	33, 22	D
Unimetrics Micro-pipettor Unimetrics Corporation 1853 Raymond Avenue Anaheim, CA 92801	5	I
Volac Universal Pipette Controller (British manufactured) Cole-Parmer Instrument Company 7425 North Oak Park Avenue Chicago, IL 60648	40	D

*PRICE RANGES

A $ 5.00 or less
B $ 1.00 - $10.00
C $ 5.01 - $10.00
D $10.01 - $20.00
E $10.01 - $25.00
F $20.01 - $30.00
G $25.01 - $50.00
H $ 30.01 - $ 50.00
I $ 50.01 - $100.00
J $100.00 or more

Figure 3, cont'd. Sources of pipetting aids

Control Practices

Labeling. All containers of chemical carcinogens should be labeled: "DANGER—CHEMICAL CARCINOGENS."

Storage/Inventory. The supervisor should maintain an inventory of stock quantities of chemical carcinogens, record the date of purchase, the disposal date (if applicable), store all materials in a specific area, and secure the area at all times. Only working quantities of chemical carcinogens should be present in the work area.

Interlaboratory Transport. If it is necessary to transport chemical carcinogens from one laboratory to another, all material should be placed in a sealed, unbreakable outer container.

Housekeeping. Housekeeping practices used in the laboratory must be capable of suppressing the formation of aerosols (6). To accomplish this, wet mopping or the use of vacuum cleaners equipped with a certified HEPA and/or charcoal filter on the exhaust should be used. Dry sweeping or dry mopping are prohibited to prevent formation of an aerosol.

Vacuum Line. Vacuum service should be protected by a disposable HEPA filter and a liquid trap to prevent entry of the chemical carcinogens into the central vacuum system, Figure 4 (8). A commercial system is available from Vacuum Guard II, Model VG201, Spectroderm International, Inc., Fairfax, Virginia 22030. If volatile chemical carcinogens are used, a separate vacuum pump or vacuum system should be used in conjunction with an appropriate laboratory-type containment cabinet.

Packaging and Shipping. Methods used for packaging and shipping etiologic agents according to 42 CFR 72.25 1972 (9), DHEW, should be adapted for the transportation of chemical carcinogens. Packages complying with these regulations were tested and proven to remain intact after receiving excessive stress beyond that normally encountered in shipping. If chemical carcinogens are physically or chemically unstable, corrosive, explosive, or flammable, the Department of Transportation regulations in 49 CFR 173 1973 (10) shall be followed. Other shipping regulations are applicable, such as Air Transport Association's (ATA's) Restricted Articles Tariff 6-D (11) and the U. S. Postal Service regulations 39 CFR 124 (12).

Decontamination/Collection. Research operations should be analyzed to determine the types of waste, quantities of carcinogenic material, and handling procedures to be employed. All chemical carcinogens, including those contained in animal carcasses, should be deactivated, degraded if feasible, or packaged

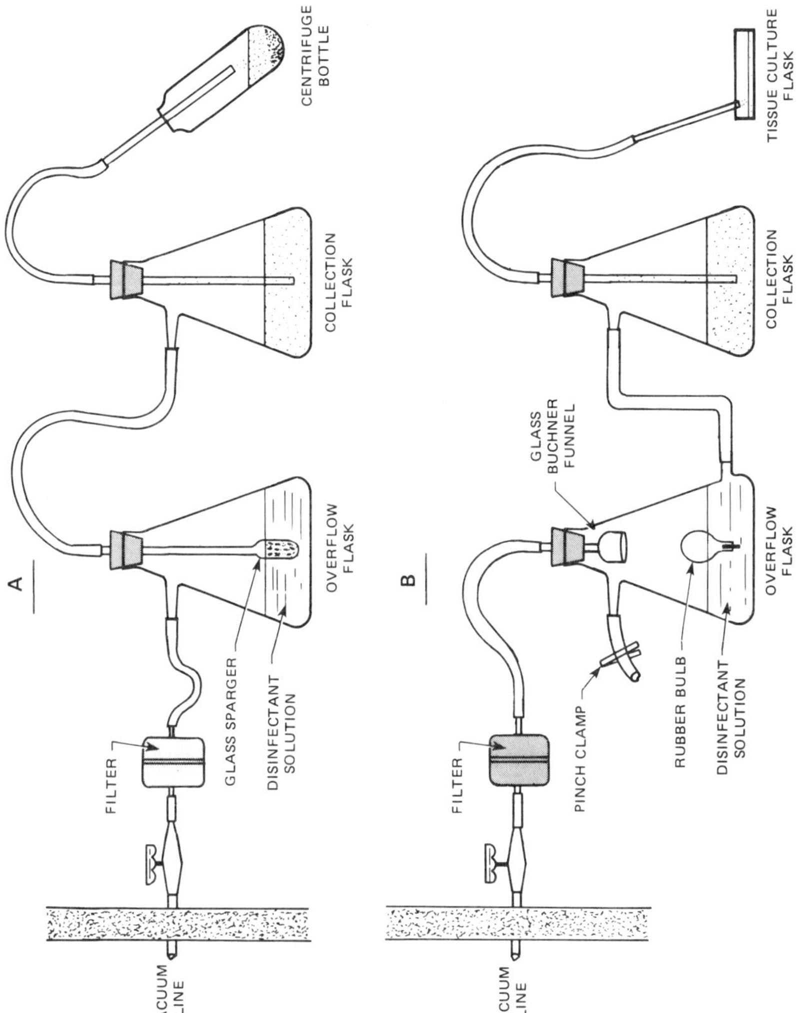

Figure 4. Two techniques for protecting vacuum system from contamination

in impermeable, sealed containers for disposal (13, 14). All discard material should be properly packaged and be compatible with other discarded chemicals before combining in common waste containers for disposal. Before they are removed, the exterior of each container is to be labeled with both the name of the carcinogen and a sign: "DANGER—CHEMICAL CARCINOGEN." Temporary storage space within the primary containment space or laboratory area should be designated for holding wastes.

Collection of chemical carcinogens, all waste products, animal carcasses and other materials from the laboratory should be performed by designated personnel on a scheduled basis or when necessary. The suspect or carcinogenic materials should not be placed on uncontrolled loading docks for pickup by general refuse collection personnel. At the National Institutes of Health, the special collection of hazardous materials, including chemical carcinogens, is done by personnel of the Fire Department. In the instance of one of the NCI contractor sites, Frederick Cancer Research Center, located at a military post, personnel from the safety offices perform this service.

Disposal. One system developed to dispose of aqueous waste containing suspect or known carcinogenic material is to place it in a burnable, one gallon plastic container filled with expandable cushioning absorbent material (K-61-Kimberly-Clark Corp. or equal, Arnold Factory Supply, Inc., 1800 S. Hanover St., Baltimore, Maryland 21230), Figure 5. The type of plastic container used must not be affected by the chemicals placed in it. As a guide, the quantity of free aqueous liquid showing in the container should not exceed about one inch in the bottom of the container. After filling, the container is placed in a double-sealed plastic bag for pickup and disposal by incineration. Contaminated flammable solvent waste may be packaged in the same manner or collected in stainless steel pressure cans for later direct injection into an approved incinerator system. If flammable solvent waste is not contaminated, it can be collected in suitable containers and recovered for reuse or sold to scavengers. All other combustible suspect or known contaminated waste can be packaged in sealed plastic bags placed in taped, durable cardboard boxes for direct transport to the incinerator. Carcinogenic material containing noncombustible material is to be degraded or deactivated, if feasible, before disposal. An alternate procedure would be to use a surface active wetting agent to physically remove the material. Then all solutions containing the carcinogenic material and cleaning materials used must be packaged and disposed of by incineration.

Specific temperatures for the incineration of all known or suspected chemical carcinogens have not been determined. Work conducted at the National Center for Toxicological Research Laboratories (NCTRL), Food and Drug Administration, Jefferson, Arkansas 72079 has shown that 2-Acetylaminofluorene may be destroyed by incineration at 1500 F measured in the secondary chamber with a minimum retention time of two seconds (personal communication with Mr. Edward J. Treglawn, Safety Officer, NCTRL). Degradation temperatures for other specific carcinogenic materials may require temperatures in excess of 1500 F for complete destruction. Although certain regulatory organizations may require the use of scrubbers on incinerators, these must be used with discretion. Scrubbers may prevent the release of hazardous material to the atmosphere, but their action deposits these materials in the waste water that then must be handled separately. If the incinerator is equipped for the injection of liquids into the firebox, similar to one developed for handling biological material (15), it may be feasible to have a closed system and eliminate the need to treat the waste water from the exhaust scrubber by yet some other method. A system for the removal of a carcinogen, 2-Acetylaminofluorene, at less than one part per billion level by non-ionic adsorption from waste water was developed for use at the NCTRL (16). If permitted by regulatory organizations, in lieu of using a scrubber to prevent the release of carcinogenic material to the atmosphere, increase the operating temperatures, increase retention time by installing internal baffles, or install an afterburner in the exhaust stack. To minimize handling by personnel of waste containers from carcinogenic laboratories, the U. S. Army, Fort Detrick developed an automatic system for dumping waste cans into an incinerator, Figure 6. Normally, containers from the carcinogenic laboratories are taken directly to the incinerator and immediately placed in a previously heated firebox of the incinerator operating at sustained temperatures.

If absolutely necessary, wastes can be transported directly to a designated burial site. The unsatisfactory results experienced in the disposal of "properly packaged" radioactive waste by burial indicate that extreme caution should be used when resorting to this method for disposal of chemical carcinogens. In January, 1976, the Environmental Protection Agency (17, 18) reported that radioactive material migrated from a surface burial facility and extended to the surrounding environment for several hundred feet from its original site. Radioactive material was detected in surface soil samples, in soil cores, in sediments from deep monitoring wells, and in sediments from intermittent streams which drained the burial sites. Again, whenever possible, if carcinogenic material cannot be rendered harmless, it should be disposed of by incineration.

Emergency Procedures

In the event of an overt accident in the laboratory, personnel should: 1) leave the area; 2) alert other personnel in the immediate vicinity; and 3) notify the safety office, fire department, medical staff, or other designated personnel who will

Figure 5. Plastic bottle with absorbent material used for the disposal of liquid carcinogenic wastes

Figure 6. Automatic incinerator charge system

respond to assist and direct the cleanup operations. If feasible, when exiting the laboratory, open flames should be turned off. If the accident occurred within a cabinet, the sash should be placed in an almost closed position, to increase the air velocity through the opening and restrict the spread of contamination. However, personnel safety is not to be jeopardized for the sake of saving experimental materials or the protection of the facility. Personnel involved directly with an overt accident should remove clothing as near as possible to the accident site in order to prevent contamination of other areas. They should immediately shower. The primary action is to minimize personnel exposure and prevent personnel injury. Before entering the accident site allow 30 minutes to elapse while the aerosolized particles settle out or are removed by the forced ventilation system. The area may be entered by personnel in appropriate protective clothing equipped with necessary cleanup material. Protective clothing, as a minimum, should consist of a complete set of disposable outer garments, shoes or boots with outer shoe covers, gloves, head cover, and appropriate respiratory protection. If feasible, the spilled carcinogenic material is to be degraded or deactivated by gently covering the spilled material with a suitable chemical. Forcible spraying of liquids is to be avoided to prevent the creation of aerosols. If the carcinogenic material spilled is aqueous, it is to be absorbed into combustible material and properly packaged for disposal. If the carcinogenic material is a solid or powder, it is to be collected and packaged for disposal.

Immediately following a cleanup, all personnel should discard protective clothing and shower. Personnel involved in cleanup operations must avoid contaminating other areas because of wearing contaminated outer clothing when passing through the building to the shower. To prevent this, double garments should be donned before entering the accident area; the outer set is left behind at the accident site when leaving after the cleanup. An alternate method is to disrobe at the accident site and put on clean clothing to walk to the shower area. All materials used in the cleanup operation are to be discarded immediately by procedures previously outlined for disposal of contaminated wastes.

Facility Design Features

Design features of facilities used for research on regulated chemical carcinogens or suspect chemicals are generally similar to those used for work with infectious microorganisms or the requirements recently promulgated for recombinant DNA research (8).

Entrances to facilities used for research with chemical carcinogens must be controlled at points of access. Access points where chemical carcinogens or suspect chemicals are used must be posted with appropriate signs: "DANGER—CHEMICAL CARCINOGEN—AUTHORIZED PERSONNEL ONLY." A shower should be available within the facility to personnel working with known or suspect chemical carcinogens. Ventilation of the laboratory area should have no recirculation of exhaust air from the work area. Exhaust air should be discharged outdoors and dispersed to the atmosphere in a manner that prevents reentrainment into the supply of the facility or of adjacent facilities. The ventilation system should be capable of maintaining directional airflow from areas of lesser to more hazardous areas. For example, air directional flow shall be from corridor to laboratory to fume hood to atmosphere.

Exhaust air from primary containment devices (fume hoods, safety cabinets or other) shall be appropriately treated by filtration using a high efficiency particulate air filter (HEPA), or by adsorption, absorption, reaction, incineration, or dilution used individually or in an appropriate combination. If HEPA or charcoal filters are used, these must be installed and operated to permit decontamination, maintenance, and replacement without exposing personnel or causing contamination of the environment. A filter system and operational procedure developed to meet these specific requirements is the Ultralok^{R} bag-in bag-out retaining system. On initial start-up, a filter is installed in the system and a plastic bag is placed on the outlet of the housing. When the filter is to be replaced, it is moved from the housing into the plastic bag. Subsequently, the bag is heat sealed. A heat seal of sufficient width will permit cutting at midpoint and removal of the filter enclosed in the bag leaving the remaining closed stub of the plastic bag on the filter housing. Then a new plastic bag containing a previously certified filter is placed over the stub and attached to the second retaining ring on the filter housing. Next, the first plastic stub is released from the first retaining ring and left in the new plastic bag. The new filter is installed in the filter housing, tested, and certified. After the airflow is measured and other operational aspects are verified to function properly, the cabinet is ready for reuse. Procedures for the certification of HEPA filters are detailed in National Sanitation Foundation Standard No. 49 (19).

Laboratory Containment Cabinets

In general, four types of cabinets are used for work with research quantities of chemical carcinogens. These are the conventional fume hood; a Class I biological safety cabinet; a Class II (Type B) biological safety cabinet; and a Class III closed glove box system (8).

Chemical Fume Hood. The chemical fume hood should have an average linear face velocity of 100 linear feet per minute (lfpm). The window sash height that gives this measured inflow should be marked on the edge wall of the cabinet. The inflow velocity to a fume hood with the sash fully opened should be 85 lfpm or more, Figure 7.

Figure 7. Chemical fume hood

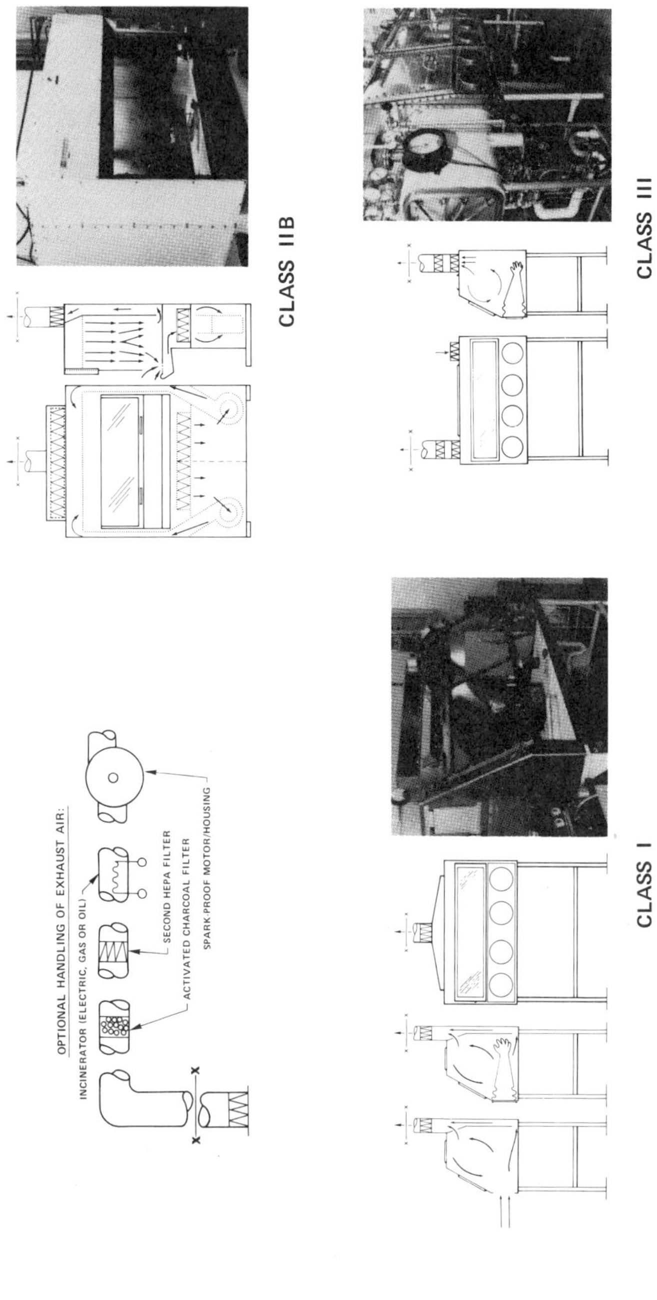

Figure 8. Biological safety cabinets and methods for handling cabinet exhaust air

Class I Cabinet. The Class I cabinet is a ventilated cabinet that offers personnel protection only. Single pass, unfiltered air from the laboratory room enters the cabinet and, after flowing across the work surface, is filtered through a HEPA and/or a charcoal filter(s) or may be incinerated before discharge to the atmosphere. The front access opening is usually fixed at eight inches. This cabinet may be used in three operational modes: with a full-width open front, with an installed front closure panel without gloves, and with an installed front closure panel equipped with arm length rubber gloves (20). If used with the full-width open front, the average inward airflow should be 100 lfpm (1), Figure 8.

Class II (Type B). The Class II (Type B) cabinet has a vertical sliding sash with a minimum work opening of eight inches and inflow velocity of 100 lfpm (±10%). It is also equipped with a HEPA filter positioned so that only filtered air enters the work zone. Approximately 30% of the air is recirculated after filtration to permit its use with research quantities of suspect or known carcinogens and tissue culture cell lines. This cabinet is not to be used with explosive or flammable solvents unless explosion-proof motors are provided. Again, the exhaust from this type of cabinet may be filtered through a HEPA and/or a charcoal filter(s) or incinerated before discharge to the atmosphere. Manifolding exhaust ducts from other Class II cabinets is permissible provided air balance is maintained, Figure 8.

Class III. The Class III cabinet system is usually of modular design, made of stainless steel, gastight, and operated under negative pressure of at least a ½ inch water gauge. It is equipped with arm length neoprene gloves. Both supply and exhaust air are HEPA filtered. Occasionally, the exhaust air may be treated before release to the atmosphere by incineration. Air ventilation is usually established at ten air changes per hour to remove contaminants from the work environment; a higher exchange rate may be necessary to control temperature rise within the cabinet below 10 F or to meet other requirements. A dunk tank or attached decontamination air lock system may be used to introduce or remove material from the Class III system, Figure 8. All primary containment devices shall be certified to meet performance criteria on initial installation, following any move or maintenance, and at least annually. Other safety features of facility design include: nonporous surfaces on floors, walls, and ceiling; self-closing doors that must remain closed to maintain air balance and directional airflows. If floor drains are present, their traps should be filled with water weekly.

Summary

Three key principles to protect employees performing work with known or suspect chemical carcinogens or other toxic substances, and to minimize or, ideally, prevent environmental contamination, include proper use of equipment, establishment of good personnel practices, and employment of safe laboratory techniques. Poor laboratory procedures cannot be offset by special equipment or facility design features. Personnel practices include protective clothing; availability of showers; mechanical pipetting devices; prohibition of eating, drinking, and smoking; and adoption of acceptable personal hygiene practices. Medical surveillance may be advisable. Operational practices must encompass a policy to:

- prevent contamination of work surfaces;
- require use of primary containment devices;
- safely package, transport and ship material;
- maintain properly secured storage areas;
- label all containers and control inventory for timely disposal of unstable materials;
- use good housekeeping practices;
- establish decontamination degradations and disposal procedures; and
- have available emergency personnel and cleanup procedures in the event of an overt accident.

The facility must be designed to control access. Signs should designate restricted areas. Clothing requirements for entry should be identified. There should be directional air control with adequate ventilation rates. Primary containment equipment such as chemical fume hoods, a Class I or Class II (Type B) biological safety cabinet, or a glove box system should be available for use with known or suspect chemical carcinogens or other toxic substances.

"Literature Cited"

1. U.S. Department of Health, Education, and Welfare, National Cancer Institute, Office of Research Safety. "Safety standards for research involving chemical carcinogens." DHEW Publication No. (NIH) 76-900. Bethesda, MD 20014. 1975.
2. U.S. Department of Labor. "Toxic and hazardous substances." Current Issue 29 Code of Federal Regulations - Section 1910.1000 - 1910.1029, Subpart Z.
3. Sansone, E.B., and Tewari, Y.B. "The permeability of laboratory gloves to selected solvents." Amer Ind Hyg Assoc J (1978) 39:169-174.

4. U.S. Department of Labor. "Personal protective equipment." Current Issue 29 Code of Federal Regulations - Part 1910, Subpart I.
5. U.S. Department of Health, Education, and Welfare, Center for Disease Control. "Cumulative supplement of NIOSH certified equipment." Atlanta, GA 30333. 1977.
6. U.S. Department of Health, Education, and Welfare, National Cancer Institute, Office of Research Safety. "Biological safety manual for research involving oncogenic viruses." DHEW Publication No. (NIH) 76-1165. Bethesda, MD 20014.
7. U.S. Department of Labor, Occupational Safety and Health Administration. "Carcinogens." Federal Register (Jan 29, 1974) 39 (20):3755-3797.
8. U.S. Department of Health, Education and Welfare, National Cancer Institute, Office of Research Safety. "Laboratory safety monograph - a supplement to the NIH guidelines for recombinant DNA research." Bethesda, MD 20014. 1978.
9. U.S. Department of Health, Education and Welfare. "Shipment of certain things." Current Issue 42 Code of Federal Regulations - Section 72.25, Subpart C.
10. U.S. Department of Transportation. "Packaging requirements for etiologic agents." Current Issue 49 Code of Federal Regulations - Section 173.387, Subpart G.
11. Air Transport Association. "Restricted articles tariff 6-D." Amendment June 25, 1977.
12. U.S. Postal Service. "Domestic mail - non-mailable matter - germs." Current Issue 39 Code of Federal Regulations - Section 124.2.
13. U.S. Department of Health, Education, and Welfare, National Institutes of Health. "Disposal of refuse: laboratory waste, dead animals, glassware and similar items." Manual Issuance 3032. NIH - 1511-1.
14. U.S. Department of Health, Education, and Welfare, National Institutes of Health. "Disposal of suspected or known chemical carcinogenic material at NIH." Manual Issuance 3032-2. NIH - 1511-1.
15. Barbeito, M.S., and Shapiro, M. "Microbiological safety evaluation of a solid and liquid pathological incinerator." J Med Primatol (1977) 6:264-273.
16. Nony, C.R., Treglawn, E.J., and Bowman, M.C. "Removal of trace levels of 2-Acetylaminofluorene (2AAF) from waste-water." The Science of the Total Environment (1975) 4:155-163. Elsevier Scientific Publishing Co, Amsterdam.
17. Environmental Health Letter. (Jan 1976) 15:2 Gershon W. Fishbein Publisher. 1097 National Press Building, Washington, DC 20045.
18. Environmental Health Letter. (September 1976) 15:17. Gershon W. Fishbein Publisher. 1097 National Press Building, Washington, DC 20045.

19. National Sanitation Foundation. "Class II (laminar flow) biohazard cabinetry." Standard No. 49. Ann Arbor, MI 48105. 1976.
20. Barbeito, M.S., and Taylor, L.A. "Containment of microbial aerosols in a microbiological safety cabinet." Appl Microbiol (1968) 16 (8):1225-1229.

RECEIVED November 22, 1978.

12

Concepts and Methodology for Toxicological Testing

B. P. McNAMARA

Department of the Army, U.S. Army Armament Research and Development Command, Chemical Systems Laboratory, Aberdeen Proving Ground, MD 21010

Numerous laws and guidelines dictate that all chemicals to which men are exposed be evaluated for short-term (ST) and long-term (LT) toxicity hazards.

The National Environmental Policy Act (Table I) states that nothing will be put into the environment which has an adverse effect on the health of man, domestic animals, wildlife, vegetation, property or cultural values.

Table I. National Environmental Policy Act of 1969 (PL 91-190)

Nothing will be put into the environment which will have short- and long-term adverse effects on:

Man
Domestic Animals
Wildlife
Property
Recreational Values
Cultural Values

The Clean Air Act Amendment (Table II) states that adverse effects on health includes the entire gamut of ST and LT toxicological effects.

Table II. Clean Air Act Amendment of 1970 (PL 91-604)

Adverse short- or long-term effects on health include:

Toxicological	Physiological
Behavioral	Teratogenic
Biochemical	Mutagenic
Immunological	Carcinogenic

The Occupational Safety and Health Act (Table III) extends these effects to include diminished health, functional capacity and life expectancy.

Table III. Occupational Safety and Health Act of 1970 (PL 91-596)

No employee will suffer diminished health, functional capacity or life expectancy as a result of the work experience.

The Toxic Substance Control Act (Table IV) states that all chemicals will be tested for their safety to man and their environment.

Table IV. Toxic Substance Control Act of 1976

Chemicals will be tested for safety to man and the environment, by the producer.

To implement these laws, guidelines (Table V) for testing (by all routes of administration for all acute, subchronic and chronic toxicological effects) have been described by the Food and Drug Administration (FDA), Environmental Protection Agency (EPA), National Institute of Occupational Safety and Health (NIOSH), Department of Transportation (DOT), Consumer Products Safety Commission (CPSC), National Cancer Institute (NCI), National Academy of Sciences (NAS), and various industrial groups. These tests appear in the Code of Federal Regulations (CFR), Federal Register (FR), and other official or quasi-official publications. These laws and guidelines indicate that a toxicological testing program should monitor the integrity of every organ or system of organs to indicate the presence or absence of ST or LT toxicological effects.

Testing a single compound for all effects requires the expenditure of hundreds of thousands of dollars and commits many research workers, many laboratory facilities, and much animal holding space for prolonged periods of time. In view of the vast number of chemicals to be tested it is necessary that more economical testing systems be adopted.

Much of the present day safety evaluation follows precepts (Table VI) which were promulgated by FDA for new drugs. This requires that therapeutic and toxic doses be studied for all toxic effects, site of action, mechanism of action, rates of absorption, distribution in the body, metabolism, excretion, time of onset and duration of effects.

Because of the vast numbers of commercial chemicals it may be necessary to base regulatory controls on much less experimental information. In some situations, it may be sufficient to demonstrate that a given dose produces "no effect". There may

Table V. Guidelines for Toxicological Testing

1. CFR 16, CPSC FED HAZARDOUS SUBST ACT. REGULATIONS, PL 92-573. OCT 27, 1972 (FR. VOL 38, NO. 187, SEPT 27, 1973)
2. PRINCIPLES AND PROCEDURES FOR EVALUATION TOXICITY OF HOUSEHOLD SUBSTANCES. NAS-NRC PUBL. 1138, 1977
3. CFR 21, FDA, JAN 1, 1972
4. CFR 46, SHIPPING, JAN 1, 1972
5. CFR 49, TRANSPORTATION, JAN 1, 1972
6. GRAZIANO TARIFF, NO. 25, APR 1972
7. CFR 40, PART 162, EPA, PESTICIDE PROGRAM
8. CFR 40, NO. 123, EPA, JUNE 25, 1975 (PESTICIDES)
9. ASSOC, F&D OFFICIALS OF U.S., 1959
10. NAS-NRC, PRINCIPLES FOR EVALUATING CHEMICALS IN THE ENVIRONMENT, 1975
11. NCI, GUIDELINES FOR CARCINOGEN BIOASSAYS IN SMALL RODENTS
12. FDA ADVISORY COMMITTEE ON PROTOCOLS FOR SAFETY EVALUATION; PANEL ON REPRODUCTION (TAP 16:264-299, 1970)
13. EPA, CRITERIA FOR EVALUATING THE MUTAGENICITY OF CHEMICALS, MAR 24, 1977
14. BARNES & DENZ, PHARMACOLOGICAL REVIEW (6:191-242, 1954) EXPERIMENTAL METHODS USED IN DETERMINING CHRONIC TOXICITY
15. WEIL & McCOLLISTER, AGR. & FOOD CHEMISTRY (11:486-91, 1963)

Table VI. Philosophies of Safety Evaluation

NEW DRUG APPLICATION

DETAILED INFORMATION ON THE NATURE OF ALL EFFICACIOUS AND TOXIC EFFECTS.

NO EFFECT LIMITS

DEMONSTRATE THAT A SPECIFIED DOSE PRODUCES NO EFFECT.

ARBITRARY CONTROLS

IN ABSENCE OF DETAILED DATA, BASE CONTROL LIMITS ON SMALLER VALUE OF ACUTE LD50/1000 OR ACCEPTED LIMIT FOR A STRONG CARCINOGEN (E.G., BIS-CHLOROMETHYL ETHER)

be situations where simple LD50 data or arbitrary control values may suffice.

In the prediction of long-term "no effect" doses there are two important concepts to consider. One is that there are predictable dose relationships between acute, subchronic and chronic toxic effects. Acute toxicity tests refers to studies wherein single or repeated doses are studied 14 days or less. Subchronic (subacute) tests refers to studies wherein the doses are given five-seven days per week for 90 days. Subchronic studies are also referred to as 13-week, three-month or short-term tests. This is in contrast to chronic, long-term, life-time tests wherein the exposure is repeated daily for two years.

The second concept is that repeated doses of any substance will produce its toxic effects (with the exception of carcinogenicity) in 90 days or not at all. Thus, the lifetime "no effect" dose can be derived from the 90-day "no effect" dose. There are indications that even carcinogenic "activity" may be detected in short-term studies.

W. Hayes (1) and C. Weil (2) *et al.*, have shown (Table VII) that there are useful relationships between LD50's, ST, LT or lifetime "no effect" doses.

Table VII. Comparison of Single Dose with 7-Day and 90-Day "No Effect" Doses (2)

LD50/6 = 7 day "no effect" dose for 75% of compounds tested

LD50/20 - 7 day "no effect" dose for 95% of compounds tested

7-day "no effect"/6 = 90-day "no effect" dose for 95% of compounds tested

LD50/120 = 90-day "no effect" dose for 95% of compounds tested

There is extensive information (Table VIII) which indicates that 1/1000 of the acute LD50 or 1/10 of the 3-month "no effect" can be given repeatedly throughout a lifetime without producing effects.

Table VIII. Prediction of Long-Term "No Effect" Doses

Dose	Repeated Doses Which Will Product "No Effect" in:
LD50/100	Three months
LD50/1000	Lifetime
3-Month "no effect" Dose/10	Lifetime

C. Weil *et al.*, (3) determined the effective and "no effect" doses of 33 compounds in rats at three months and two years (LT) (Table IX). These compounds were diverse in chemical structure, pharmacologic type, and toxicity as judged by LD50's or ST and LT "no effect" doses.

Table IX. Summary of Information of Weil et al. (3)

33 Compounds	Agriculture chemicals, pesticides, veterinary products, thickeners, stabilizers, additives, antimycotics, water treatment, food packaging materials
Species	Rat and dog
Dose Ranges	
LD50	17 - 31,000 mg/kg
Short-term "no effect" dose	2.25 - 1,750 mg/kg
Long-term "no effect" dose	2.25 - 3,750 mg/kg

In the above study animals were observed for 36 criteria of toxicity (Table X) including mortality, food intake, weight, pathology, hematology, blood chemistry, CNS effects, fertility, cholinesterase, and neoplasia. The most sensitive criteria were body weight gain, liver and kidney weight ratios, and liver and kidney pathology. These investigators believe that only these parameters need be followed in the 13-week test. It was their opinion that the rat was a more sensitive indicator of the "no effect" level than the dog.

Table X. Summary of Information of Weil et al. (3)

36 Toxicity Criteria Including:
Mortality, food intake, weight, pathology, hematology, blood chemistry, central nervous system, fertility

Most Effective Criteria:
Body weight gain, liver and kidney weight/body weight, liver and kidney pathology

The safety factors for the 33 compounds in rats are shown in Table XI. This table shows that: (1) for 27% of the compounds the LT "no effect" dose was the same as, or greater than (tolerance) the ST "no effect" dose; (2) for 51% of the compounds the LT "no effect" was less than the ST "no effect" dose by a factor of only 2 or less and (3) the ST "no effect" dose divided by factors of 5, 10 or 12 would encompass the LT "no effect" dose for 91, 97 and 100% of the compounds, respectively.

Table XI. Safety Factors for 33 Compounds in Rats (3)

Factors*	No. of compounds influenced	Percent of compounds influenced
1	9	27
2	17	51
5	30	91
10	32	97
12	33	100

*Short-term no effect dose/factor=long term no effect dose

The important feature of these data is that there is a 27% chance that the ST given repeatedly will not produce toxic effects in a lifetime and a 95% to 100% likelihood that 1/10 to 1/12 of the ST "no effect" dose can be given repeatedly throughout a lifetime without producing toxic effects.

To test the concepts advanced by Weil et al., 82 studies covering 122 compounds and 566 dose levels were collected from the literature. These studies were analyzed for effects obtained at 13 weeks and at two years. The analysis is shown in Table XII.

Table XII. Summary of Data from Various Literature Sources

Number of studies	82	
Number of compounds	122	
Number of doses*	566	
	No. of compounds/doses*	Percent
No positive doses at three months but some positive doses at two years.	3/122	2.5
Some positive doses at three months but more positive doses at two years.	5/122	4.1
All doses which were "no effect" at three months were also "no effect" at two years.	114/122	93.4
Doses* which were "no effect" at three months and two years.	551*/566*	97.4

The studies of Weil et al., were conducted by the same investigators using a relatively homogenous group of rats and consistant criteria for toxicological measurements. The "no effect" doses were determined with precision.

The literature review included compounds, investigators, animal species, and criteria of toxicity which varied greatly. Despite the welter of information, an agreement with the data of

Weil et al., concerning relationships of LD50's, ST, and LT "no effect" dose for individual compounds could be seen.

J.P. Frawley[4] collected two-year toxicity data on 220 substances. A distribution of the "no effect" levels is shown in Table XIII.

Table XIII. Summary of Distribution of the "No Effect" Level

"No effect" Level (PPM)	All compounds (220)	Heavy Metals & Pesticides (88)	Other (132)
<1	5	5	0
<10	19	19	0
<100	40	39	1
<1000	101	72	29
<10,000	151	86	65

Only 19/220 compounds showed any toxic effect below 10 ppm; all 19 were pesticides or heavy metals. Only 1/132 other compounds was toxic below 100 ppm. Frawley implied that a safe level of 0.1 ppm could be established without experiment for all compounds with possibly the exception of pesticides and heavy metals.

All of the preceding indicates that commercial chemicals might be controlled on the basis of exposure risks and relatively limited toxicological data (Table XIV).

Table XIV. Regulatory Controls and Costs for Commercial Chemicals Based on Readily Available Information and Predicted Risks

Risk Prediction	Control Requirements	Very Rough Cost ($K) Oral	Very Rough Cost ($K) Inhalation
No information	Corrosivity LD50/1000 Established limit of known potent carcinogen	3	5
Moderate	13-week test (abbreviated) Mutagenesis, (Ames, micronucleus, lymphoma, Drosophila) DNA repair, cell transformation, one generation reproduction	45	80
High	13-wk test (complete toxicology Mutagenesis, DNA repair, cell transformation, 2nd & 3rd generation reproduction tests NCI cancer test	375	500

For a new compound when there is no information upon which to predict risks or toxicity, a level of the acute LD50/1,000 might serve as a preliminary control standard for most toxicological effects. For carcinogenicity, a worse case could be assumed, and the limit would be that of a potent known carcinogen, e.g., bis-chloromethyl ether. The lower value (toxicity or carcinogenicity) would be the limit.

The initial limit based arbitrarily upon a known carcinogen could be relaxed upon submission of favorable short-term testing data for carcinogenicity and mutagenicity (Ames test, cell transformation, DNA repair, chromosome damage, Drosophila etc.,) The test compound could be compared to a known carcinogen (and mutagen). The new limit for the test substance would be decreased in proportion to its potency as related to the standard (e.g., bis-chloromethyl ether).

The LD50/1000 limit could be ameliorated by consultation with EPA and submission of favorable data from an abbreviated 13-wk test. A one-generation reproduction study may be desirable also. A more detailed 13-wk test might be submitted by the manufacturer to further relax the control values. For substances of great risk (quantity produced, population exposed, toxicity, chemical structure) a more detailed 13-wk test in which all organs and functions are monitored may be requested by the control agency for protection of public health. This could include short-term tests and long-term tests for reproduction and carcinogenicity.

Current laws and guidelines dictate that chemicals be tested for all ST and LT effects on all body organs and functions. A minimal type program designed to comply with the laws and guidelines will be described. On most of the figures the drawings illustrate the test and the caption shows the organs or functions being tested. The program progresses from simple ST tests to the more complex and expensive LT tests.

Usually the first test is a screen in small rodents (Figure 1) to reveal the presence or absence of signs over a wide range of doses. The compound is given by the appropriate route and appearance of signs is noted periodically using a checklist. Dose-response regression lines are developed for the various signs. The figure shows the dose, hypothetical effects and percent of population showing each effect.

No single species is a satisfactory indicator for all qualitative and quantitative effects of a compound in man. The substance is given to various species by appropriate routes. If the pattern of regression lines is similar for a variety of mammalian species, there is reason to assume a similar pattern would apply to man. However, species differences usually are found.

At this time, it is desirable to obtain some information on metabolism (Figure 2). It is assumed that species which most resemble man in metabolism will be most like man in a toxicological response to the substance.

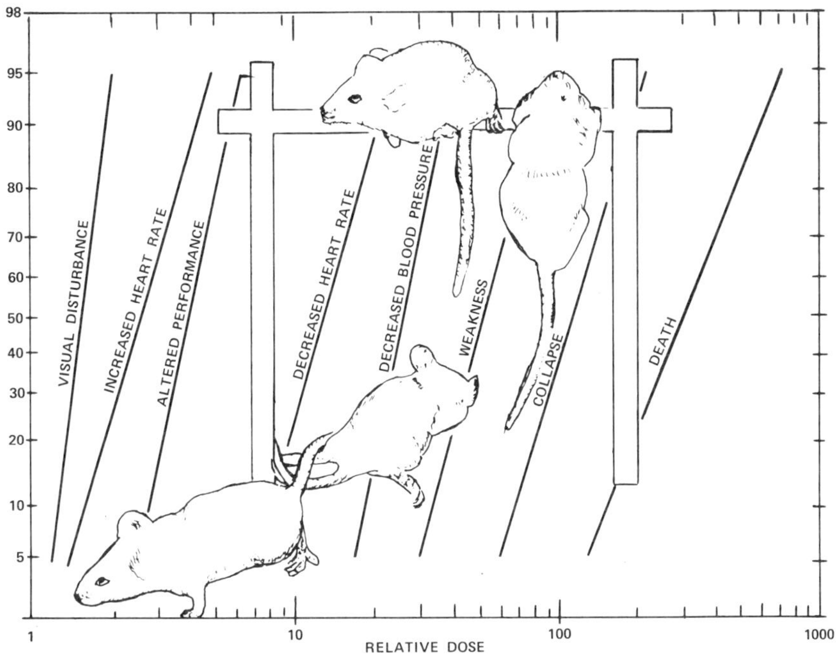

Figure 1. General toxicological effects. Mouse screening and toxicological regression lines for a hypothetical compound.

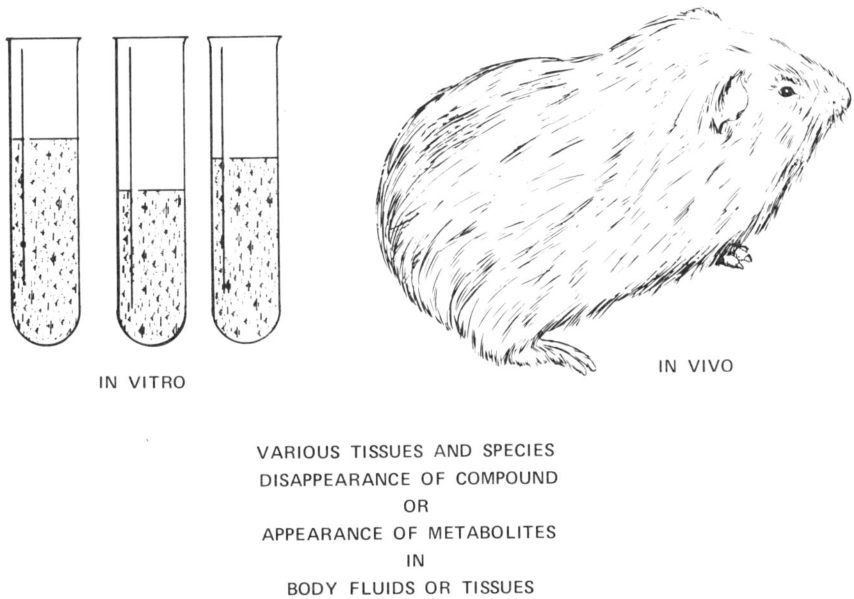

Figure 2. Metabolism, detoxication, liver function, kidney function, excretory processes

Tissues from various species, including man can be compared for their ability to metabolize a substance, "in vitro" and "in vivo." This usually is done by noting the disappearance of the material from the tissue or the appearance of a metabolite.

There are several tests (Figure 3) which have been required under the Hazardous Substance Labeling Act and previously by FDA and DOT and now by CPSC.

These tests usually are performed at an early stage. They are relatively simple. They give guidance for safe handling and shipping. They furnish comparative information on various routes of administration. These tests show damage to eyes and skin; and lethality following administration by the percutaneous, oral or inhalation routes.

In addition to the local eye test (Draize Test) just mentioned, the lens and retina should be examined (Table XV).

Table XV. Toxicological Effects of Chemicals on Vision

Local	-	Draize Method
Lens	-	Opthalmoscope or slit-lamp bimicroscope
Retina	-	Ophthalmoscope, electroretinograms, histology

Hematology, blood chemistry and pathology provide important toxicological information (Figure 4). Blood cells and chemicals are measured before and periodically throughout the experiment. Tissues from animals which die or are sacrificed are examined grossly and microscopically. These studies reveal toxic effects on all organs and functions shown.

Adverse effects on the reticuloendothelial system (RES) can be shown by impaired ability to produce antibody protein or by the development of hypersensitivity.

The ability to produce globulins and cellular proteins can be noted in the hematological studies.

The Landsteiner technique in guinea pigs (Figure 5) is a routine test for sensitization. The test substance is applied to the skin three times a week for three weeks. After a two-wk rest period, a previously non-effective dose is applied. The appearance of erythema, edema or necrosis indicates sensitization.

Learned performance tests (Figures 6 and 7) reveal deficits in the mental and physical abilities of animals.

At the present time we use a battery of three learning and memory tests. These are (1) Unlearned behavior is measured by activity; (2) Ability for new learning is determined by comparing experimental and control animals on their time to learn to respond to a signal and retreat to a safe area thereby avoiding an electric shock; and (3) Memory (or old learning) is measured by ability to press four levers in a given sequence for a receipt of a food reward.

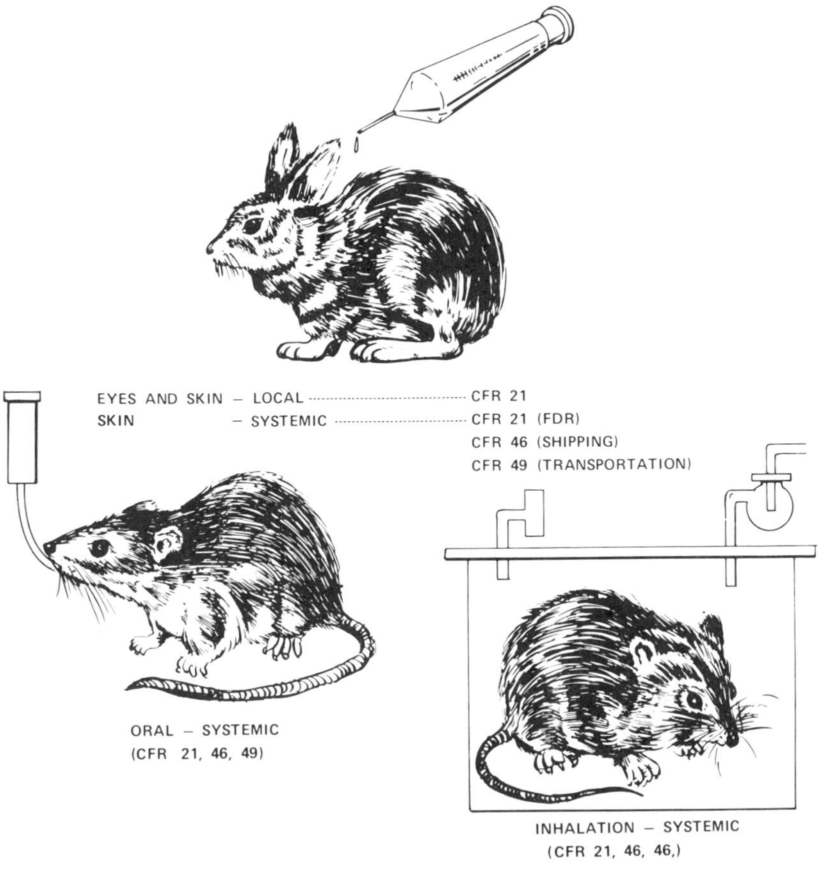

Figure 3. Local effects on eyes and skin lethality—oral, percutaneous, inhalation routes

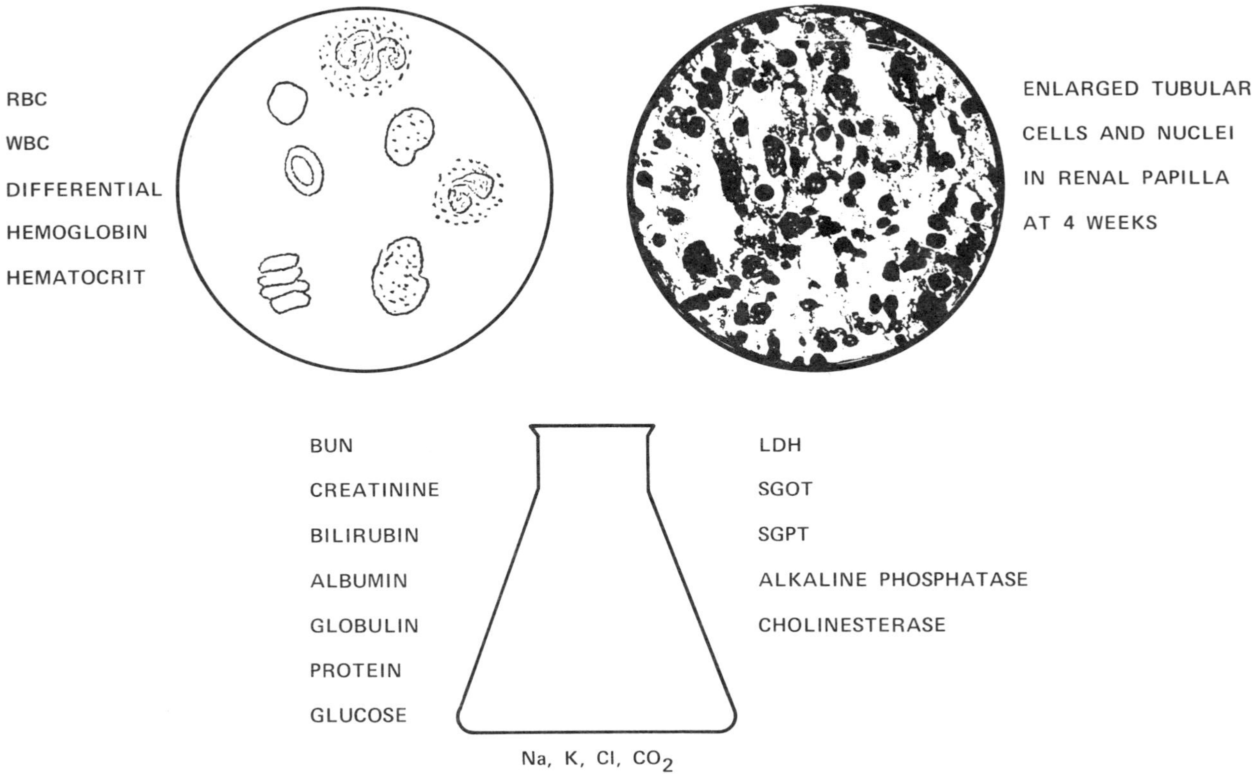

Figure 4. Hematopoietic system, liver and kidney function, endocrine system (salt–H_2O) reticuloendothelial system, all organs, neoplasia

Figure 5. Reticuloendothelial system. Sensitization in guinea pigs. Antibody and other protein synthesis in various animal species.

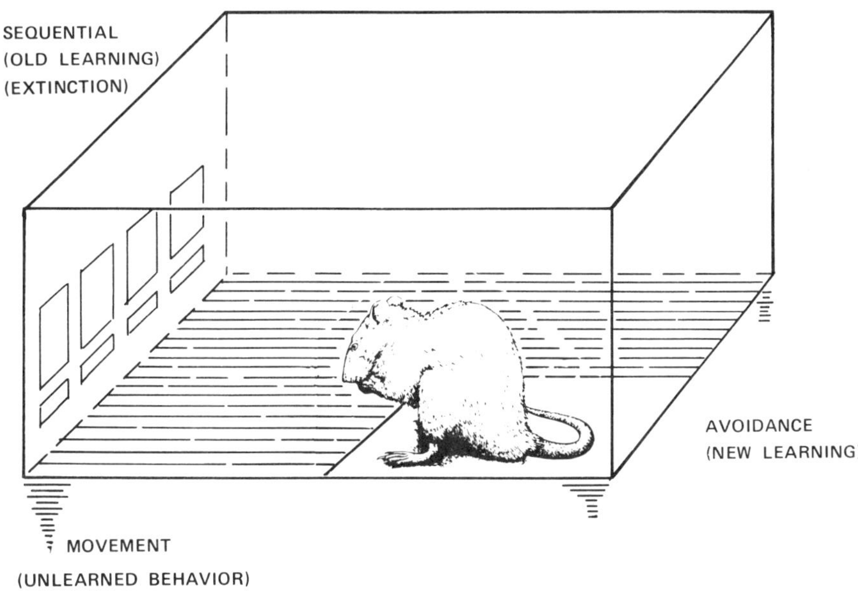

Figure 6. Cerebral function (learning and memory)

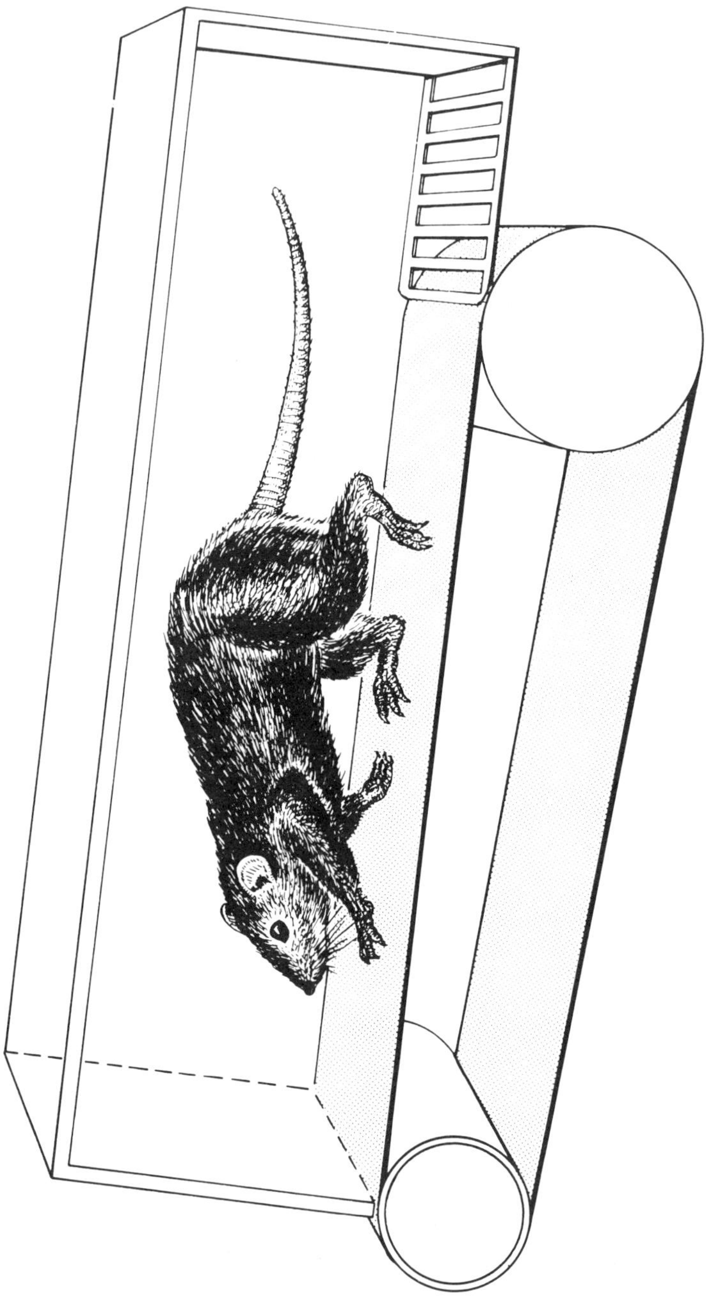

Figure 7. Physical performance

Defects in physical performance can be noted in rats running on a small treadmill. Chemicals may have little effect on resting animals but profound effects on animals while running.

The measurement of blood pressure (BP), heart rate (HR), electrocardiogram (ECG), blood flow, neuromuscular function and respiratory rate and amplitude can be used to study all of the systems shown in the caption in Figure 8.

In addition to the cardiovascular, the neuromuscular, and respiratory systems, these measurements can be used to reveal abnormalities in many sensory receptors, neural pathways and effector organs throughout the body.

As shown, tetanizing stimulation of the sciatic nerve produced contraction of the gastroneumius muscle and reflex increase in BP and respiration. Occlusion of the carotid artery causes increased BP and HR. Pressure on the ocular muscles decreases BP and respiration. Due to their transitory nature, many other reflexes can be used.

Numerous tests are used to study the various aspects of reproduction. The FDA three-generation test in rats (Figure 9) reveal dominant lethal mutations in their first generation and recessive mutations in the second and third generation. It is unlikely that the second or third generation will supply additional information on mutagenicity. Teratology must be studied separately--or as an add-on to the FDA test. A three-generation test requires about 10 months of working time.

It is possible to perform a one-generation reproduction test (Table XVI) which would assess the impact of all ST and LT toxic effects on mating, fertility, fetal toxicity, dominant lethal mutagenesis, teratogenesis, gestational and post-natal effects on survival, growth, lactation, etc., in a period of four to five months.

Table XVI. Effects Measurable in a One-Generation Reproduction Test

1. Mating	4. Dominant Lethal Mutation
2. Fetal Toxicity	5. Teratogenesis
3. Fertility	6. Gestational Effects
7. Post-Natal Effects on Survival, Growth and Lactation	

Mutagenesis, generally, is revealed in microorganisms, molds, plants, cell cultures, or insects. These tests reveal mutagenic potential but not mutagenic hazard to man. An important value of such tests is to compare the potency of new chemicals with "known" mutagens.

The Department of Health, Education and Welfare Committee to Coordinate Toxicology and Related Programs recently (Apr 77) advocated a battery of short-term tests (Figure 10) which would demonstrate point mutations, DNA damage, and chromosome

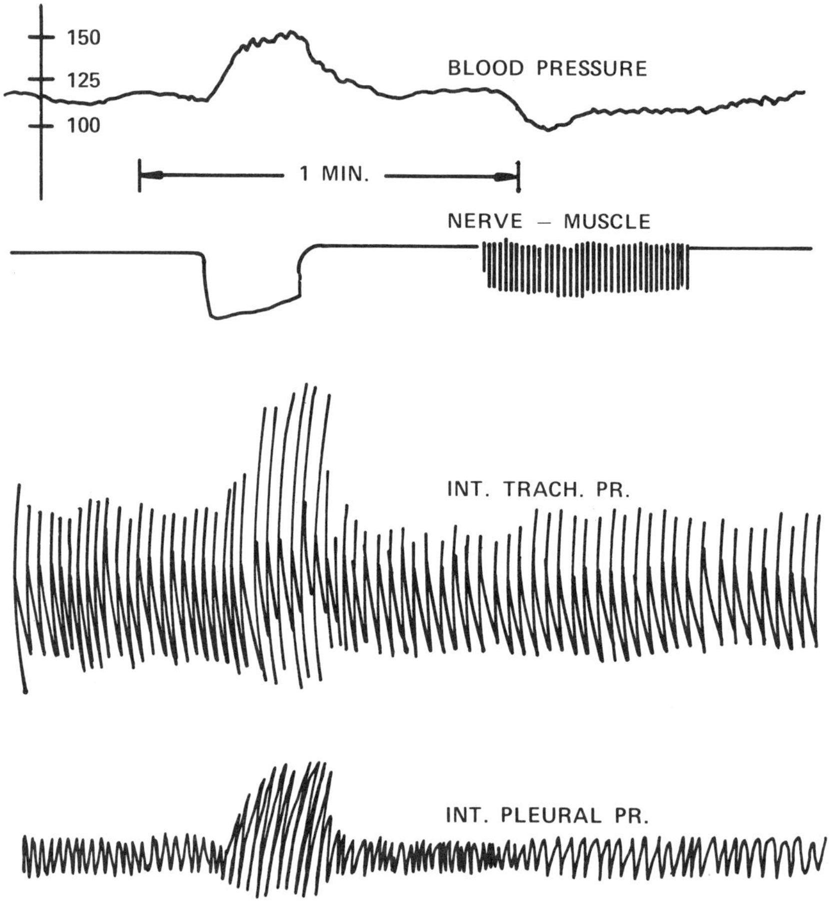

Figure 8. Assessment of central, autonomic, neuromuscular, cardiovascular, respiratory, and neurohumoral receptor and effector systems

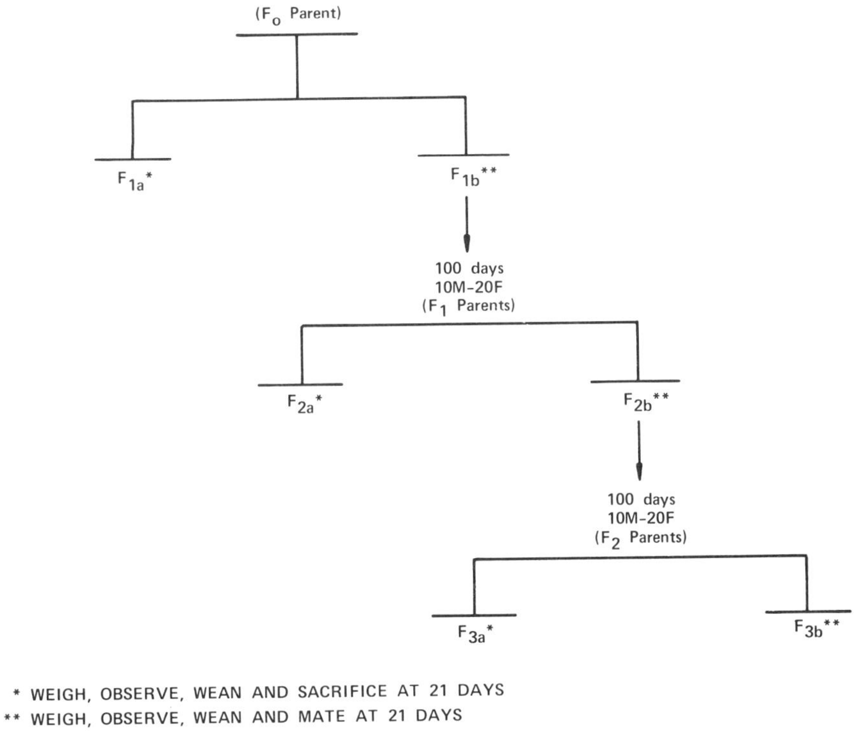

Figure 9. FDA multigeneration reproduction study in rats

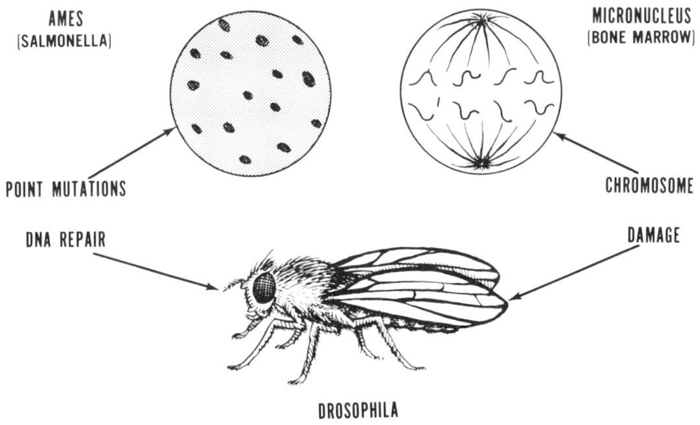

Figure 10. Mutagenicity. Possible short-term tests.

abnormalities. Included were the Ames Test, Mammalian Lymphoma Test, Micronucleus Test, Drosophila Test and others.

Mutagenesis often is considered to be a step in carcinogenesis (Figure 11). There is good correlation between mutagenic and carcinogenic potency for a number of compounds. A number of short-term tests for carcinogenicity is being considered also.

The carcinogen bioassay method advocated by the NCI tests mice and rats for a period of 24 months.

Short-term tests are receiving favorable consideration in many laboratories. The report of Stoltz _et al._,[(5)] and other experts in the field conclude that tests involving mutagenicity (point mutations, DNA damage, chromosome changes as shown by Ames, micronucleus, mouse lymphoma, Drosophila tests), DNA repair synthesis and cell transformation offer promise as screens for carcinogens.

Figure 11 illustrates the wild, disorganized growth of transformed cells and radiogram showing increased DNA repair synthesis. Mutagenesis was mentioned previously.

The literature is replete with pharmacological, histological and biochemical effects produced by known carcinogens. Many of the effects appear within a few hours (or days) and disappear within several weeks of exposure. Many investigators feel that such indicators can be expected generally. Table XVII addresses precancerous signs.

Table XVII. Precancerous Indicators

Clinical signs or measurements:
anorexia, vomiting, hemorrhage, decreased body weight, hyperbilirubinemia, BSP retention, decrease in hemoglobin, platelets, leukocytes and neutrophils

Histological:

Organ damage -	liver, lungs, kidney, breast, gonad, bone marrow, intestinal epithelium
Tissue changes -	proliferation, hyperplasia, vesiculation of endoplasmic reticulum, enlarged nuclei, squamous metaplasia, mitosis, fibrosis, necrosis

Biochemical and enzymatic:
Changes in glucose-6-phosphatase, forminoglutamic acid transferase, urocanase, formylase, methyl-H_4-folate dehydrogenase, histidiase, glutamic oxalacetic transaminase and carbonyl transferase and ornithine

At the present the guidelines for long-term mammalian carcinogenesis recommend a 24-month test in small rodents (Figure 12). It is advisable to use male and female animals in each group. Controls should include: (1) normal controls for

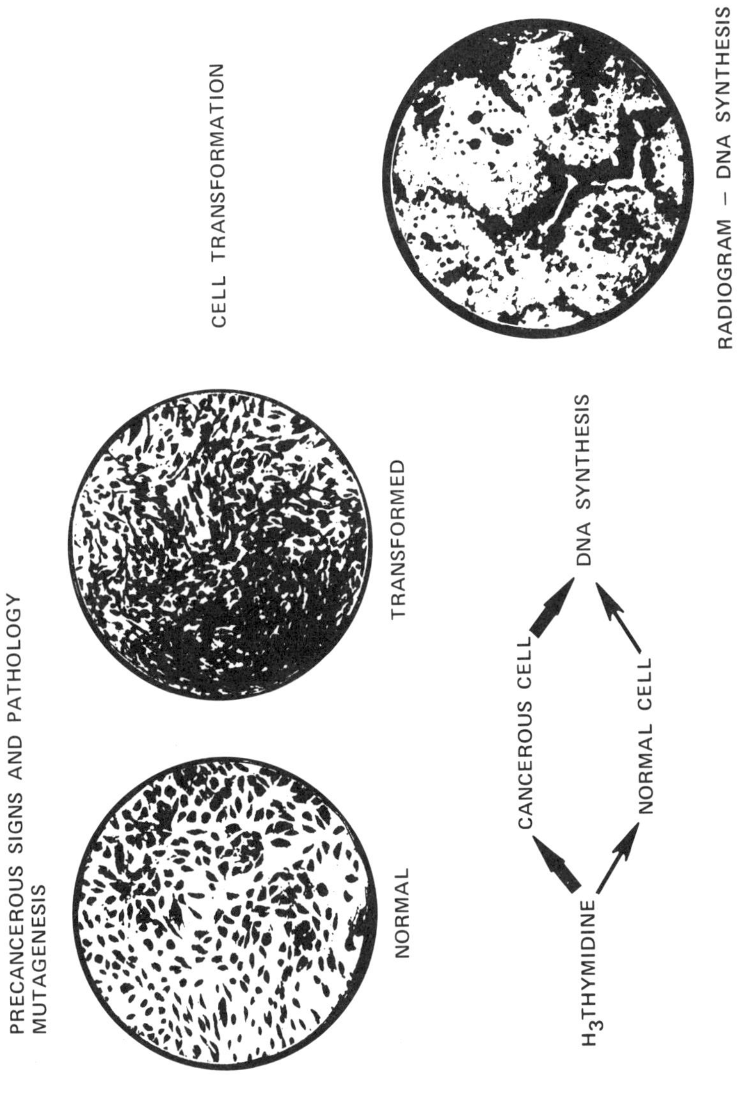

Figure 11. Carcinogenicity. (Possible short-term tests).

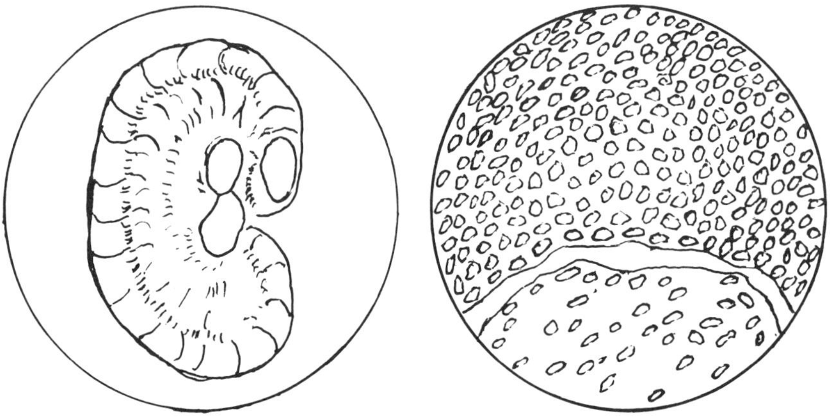

THE TEST

MALE AND FEMALE MICE AND RATS FOR EACH MATERIAL

VEHICLE CONTROLS — TEST MATERIAL DOSE 1
POSITIVE CONTROLS — TEST MATERIAL DOSE 2
NORMAL CONTROLS

Figure 12. Carcinogensis. Tumors of the kidneys.

Table XVIII. General Plan for Toxicological Studies

TESTS	ORGAN, SYSTEM, FUNCTION, OR TOXICOLOGICAL PROCESS MONITORED
ACUTE LD50	
TOXIC SIGNS	GENERAL, ALL SYSTEMS
SHORT-TERM TEST 2 DOSE LEVELS	
DAILY TOXIC SIGNS	GENERAL, ALL SYSTEMS
BLOOD CHEMISTRY & HEMATOLOGY	HEMATOPOIETIC SYSTEM (RBC-WBC), ANTIBODIES, PROTEIN, SYNTHESIS, LIVER AND KIDNEY FUNCTION, ENDOCRINE SYSTEMS, SALT-WATER BALANCE
PATHOLOGY	BRAIN, HEART, LUNG, KIDNEY, LIVER, INTESTINES, MITOSIS, SPLEEN, BONE MARROW, GONAD, LYMPHOID AND MYELOID TISSUE
EYE AND SKIN	LOCAL
EYE EXAMINATION	VISION
PHYSIOLOGY	CENTRAL NERVOUS, CARDIOVASCULAR, RESPIRATORY, AUTONOMIC NEUROMUSCULAR SYSTEMS, REFLEXES
PERFORMANCE TESTS	MENTAL & PHYSICAL FUNCTION
LANDSTEINER (GUINEA PIG)	RETICULOENDOTHELIAL SYSTEM – SENSITIZATION
AMES & MICRONUCLEUS TESTS	
DROSOPHILA	MUTAGENESIS, CARCINOGENESIS
REPRODUCTIVE SCREEN (1 & 3 GENERATION)	REPRODUCTION, MUTAGENESIS, TERATOGENESIS, FERTILITY AND FETAL TOXICITY
NCI CANCER TEST (RATS)	CARCINOGENICITY

spontaneous tumors; (2) controls for tumors due to the solvents; and (3) positive controls to show that tumors can be produced in the species under the experimental conditions. The nature of the lesion must be verified by histological examination.

In summary, the performance of all of these tests should at least monitor all ST and LT toxicological effects of the test substance (Table XVIII).

The time required is at least three years and the cost may reach or exceed $1,000,000. Hopefully future legislation may ease the burden for toxicological testing of commercial chemicals.

LITERATURE CITED

1. Hayes, W., Jr. "Essays on Toxicology", Vol. 3:65-67, Academic Press. New York. 1972.

2. Weil, C.S., Woodside, M C., Bernard, J.R. and Carpenter, C.P. Toxicol. Appl. Pharmacol. (1969) 14:426-431. Single Peroral One-Week and Ninety-Day Rat Feeding Studies.

3. Weil, C.S., and McCollister, D.D. Agr & Food Chem. (1963) 11:486-491. Safety Evaluation of Chemicals. Relationship Between Short-Term and Long-Term Feeding Study in Designing an Effective Toxicity Testing.

4. Frawley, J.P. Food Cosmet. Toxicol. (1967). 5:293-308. Scientific Evidence and Common Sense as a Basis for Food Packaging Regulations.

5. Stoltz, D.F., Poirer, L.A., Irving, C.C., Stitch, H.F., Weisburger, J.H. and Grice, H.C. Toxicol. Appl. Pharmacol. (1974) 29:157-180. Evaluation of Short-Term Tests for Carcinogenicity.

RECEIVED December 14, 1978.

13

Department of Defense Chemical Ammunition Safety Program

R. A. SCOTT, JR.

Department of Defense Explosives Safety Board, 2461 Eisenhower Avenue, Alexandria, VA 22331

After World War I, ammunition and explosives which had been returned from combat areas accumulated in United States storage depots which were inadequate for the safe storage of such large quantities. Many incidents, fires, and explosions occurred involving these stores. House Document 199, Ammunition Storage Conditions, became the foundation for much of the planned development of explosives safety standards as we know them today. In 1928, Congress recommended and passed an appropriation act which provided for the establishment of a permanent joint board to handle explosives safety matters. It was determined that the mission of the board should be (A) to advise and confer with the military departments, (B) to keep advised of the storage of ammunition and ammunition components to assure proper separation, and to prevent hazardous conditions from developing which could endanger life and property within and without storage reservations, (C) to study and approve plans for new storage facilities prior to construction or when required, to make recommended changes needed to satisfy safety standards, and (D) to hold both regular and special meetings. The original joint Army-Navy board evolved into the present Department of Defense Explosives Safety Board, DDESB, chartered by DoD Directive 5154.4.

Today, DDESB is concerned with the same explosives safety aspects of munitions manufacture, storage, transportation, and disposal as was recommended by Congress in 1928. An additional functional area added in 1968 by the Secretary of Defense is the establishment of chemical safety standards and a chemical safety program for chemical agents and components of chemical ammunition.

A brief resume of the chemical safety functions and responsibilities of the DDESB and some of the policies and philosophies will be provided herein.

The DDESB chemical ammunition safety program establishes a uniform Department of Defense approach for the development of safety programs for chemical agents and associated weapon systems, either adopted as chemical warfare items by the military departments or being investigated for such use, which, through chemical properties produce incapacitation or death.

The program also provides for the establishment of safety standards to cover the entire life cycle of chemical weapons. These safety standards include medical health standards applicable in the protection of health of those actually or potentially exposed to chemical agents.

Prior to initiation of chemical ammunition operations, the military departments must obtain DDESB approval for site plans and safety submissions for proposed operations. Site plans are required for construction or modification of ammunition facilities to be located adjacent thereto.

DDESB policies, philosophy, and safety criteria are reflected in published chemical ammunition standards which were published in the Federal Register (41 FR 20686) and in DoD Ammunition and Explosives Safety Standards (1). These standards are used to review the site plans, safety submissions, and in conducting worldwide safety surveys of activities that store or handle DoD chemical ammunition. The surveys, however, are not conducted just to evaluate compliance with chemical ammunition and explosives safety standards. The surveys are a means whereby the DDESB Secretariat can keep informed on the military departments' safety posture, determine the need for new standards or change to existing ones, and detect hazardous conditions.

The DDESB evaluation process is initiated by review of toxicological agent hazards associated with the operation. For this purpose quantitative information is used to verify the adequacy of control limits in preventing adverse effects when members of a specified population are exposed to an agent for a specified time period.

The population to be effected by the unexpected release of agents is calculated using DDESB quantity-distance standards based upon providing hazard radii from the source (2). The DDESB model for chemical hazard prediction provides for limiting the vertical expansion of the agent cloud to the surface mixing layer and also provides criteria to accommodate time-space variations of meteorological and other environmental factors. DDESB has funded the Naval Surface Weapons Center to develop a graphic chemical hazards sliderule using the T.P. 10 methodology for simplified use in deriving the downwind hazard distances.

This same model is used to calculate the maximum credible event (MCE) which is defined as that unintended, unplanned, or accidental adverse occurrence which causes release of agent from an ammunition item, bulk container, or process. It must be realistic with a reasonable probability of occurrence. It is necessary to hypothesize an MCE to enable calculation of the magnitude of a worst case hazard. The hypothetical MCE for any given situation will be based upon the nature and characteristics of the agent involved, ammunition, container, configuration, and location. The number of ammunition items involved, the quantity of agent released, and the percentage of agent disseminated shall be assumed, based upon available test data or, in the absence of data, the best technical judgment available. Such MCE's are evaluated and the worst case hazard source strength shall be selected, based on the type and quantity of agent released and type of release; e.g., by aerosolization or evaporation for explosive or non-explosive dissemination, respectively.

The public access exclusion distance (PAED) provides protection analogous to the inhabited-building distance for explosives. The hazard PAED zone calculated from the MCE shall represent that arc from the agent source containing no more than 10.0, 4.3, and 150.0 $mg\text{-}min/m^3$ of GB, VX, or HD, respectively. Positive means shall be taken to assure that no persons, not directly associated with chemical weapons operations, enter areas so defined.

In the review of site and safety submissions for agent operations, a means must be provided to assure the agent concentrations are quantitatively documented by agent detectors. These detectors are needed to verify on a real-time basis that concentrations of released chemical agents will be detected prior to exceeding the control limits during the operations.

Facility requirements for agent operations are evaluated on a case by case basis. In general, the ventilation system must be designed so that the atmospheric pressures within facilities will be maintained at a progressively enhanced negative pressure below that of surrounding areas as one approaches agent work areas. When entering agent contaminated operational areas, isolation of clean areas from those containing released agent is assured by means of this negative pressure and by the use of air locks. Room surfaces in agent work areas must also be treated so that surfaces can be readily decontaminated. Agent hoods must be calibrated to provide a face velocity through the working opening which is satisfactory for the total containment of agents within the hood structure. Air from agent facilities must be directed to the outside of the facility and be filtered or undergo other air treatment to remove all agent in the effluent. Another facility feature needed for agent work is a dedicated agent sump which permits retention of liquid wastes for decontamination purposes.

Certain operations are inherently hazardous and appropriate containment of the various hazards involved is necessary for the protection of the employees performing such work, the protection of other employees at the installation who are not associated with such work, and the protection of the general public outside the installation. Personnel responsible for planning, designing, and accomplishing such operations must assure that adequate safety is provided by incorporating the appropriate types of hazard containment. The various circumstances and facilities that may be encountered with such operations prevent pre-definition of specific detailed containment requirements for each agent, each ammunition item, and each operation. Nevertheless, the general principles of hazard containment set forth below will be normally incorporated in operations such as manufacture, disassembly, demilitarization, and disposal.

There are two types of containment: "total containment," and "vapor containment." Irrespective of which type of containment is provided, the containment structure or facility will be equipped with a means of entrapping or detoxifying the evaporated or aerosolized chemical agent by filters, scrubbers, incinerators, or other appropriate means. Total containment and vapor containment are described as follows:

Total containment requires a facility designed and tested to be of sufficient capacity and strength to contain combustion or detonation gases, fragments, and agent from the largest explosion that could occur, based upon the propagation characteristics of the ammunition. Currently, there are two basic designs for such total containment. One design consists of a chamber capable of retaining all of the fragments and explosion effects, and preventing release of detectable quantities of agent. The other potential design consists of a suppressive shield capable of retaining the fragments and sufficiently attenuating the blast forces while a chamber retains the combustion gases and prevents the release of agent. Further research and development are required on this concept design for use in chemical demilitarization operation.

Vapor containment will consist of a facility designed to provide negative pressure, controlled air flow, and walled or multiple walled enclosures which will contain any detectable quantities of agent released.

Designs for vapor containment are usually tailored to the operation involved.

Containment is not required for operations associated with storage activities. In the event of a leaker, measures such as personal protective clothing and equipment are mandatory to protect operating personnel during the application of decontamination procedures to neutralize the escaping agent, and to repair, encapsulate, or transfer agent from the leaking ammunition or container.

The selection of the type of containment is dependent upon the nature of the operation involved. Total containment is required for those operations involving ammunition which contain explosive components as well as toxic agents, whenever the operation may subject the explosive components to a potential initiating stimulus. Vapor containment is required for those operations involving toxic agents in bulk or in ammunition without explosives components; and for those operations involving ammunition containing both toxic agent and explosive components wherein the operation does not subject the explosive components to a potential initiating stimulus.

Many of the chemical safety functions and responsibilities of the DDESB and many of the policies and philosophies are reflected in the design features of the Army's Chemical Agent Munitions Disposal System (CAMDS) under construction at Tooele Army Depot, Utah. Some of the design features of this facility are indicated below:

A. The regulation of the agent quantity in CAMDS OPERATIONS assures that any released agent is maintained on Tooele Army Depot property during a maximum credible accident.

B. The CAMDS plan involves disposal of mustard agent by incineration processes.

C. GB and VX will be chemically detoxified by reaction with sodium hydroxide solution or a chlorinalysis process, respectively.

D. Explosives and energetic materials incorporated in the munitions will be destroyed by incineration.

E. Inert components and metal parts will be mechanically demilitarized and thermally decontaminated.

F. The amount of agent being processed in any one operation is of a sufficiently small quantity as to be completely contained and absorbed by a network of filter scrubber systems. Further, the facility is designed to isolate one activity from another in an independent and completely autonomous manner by numerous air changes per hour in toxic areas (up to 25/hr.).

G. Throughout the demilitarization operations a network of detectors will provide accurate and immediate indications of agent contamination. If an area of contamination is detected by alarm systems, that area will be chemically decontaminated and checked to verify that there is no detectable concentration of agent present.

H. Eight perimeter stations around the CAMDS site monitor the atmospheric effluent constituents to assure that air pollution limiting values for each effluent are not exceeded.

Impeccable safety has been a primary goal of the CAMDS design since inception. In evidence of this, prior to introduction of munitions (possibly unsafe-leaking or ruptured) into the CAMDS area, operational procedures require inspection at the storage igloo; again after transport to the CAMDS site, and before entering the unpack area.

The munition route following the unpack area is the Explosive Containment Cubicle (ECC) containing four different demilitarization machines which will be remotely operated. Operation within the ECC will be followed by closed circuit television surveillance. Since there is a chance of an explosive fire incident, the ECC is designed for explosive and agent containment. Each step in demilitarization of munitions is controlled by a computer which also stops the operations within the ECC when a problem develops or equipment maintenance is needed.

Advancement of the munitions in the demil operations by conveyors to the furnaces and explosive agent destruction operations have been reviewed with possible failure modes, fire, toxic release, and deterrent actions.

In summary, an assessment of the sequence of events which

comprise the process flow for any given munition being demilitarized in the CAMDS reveals an operational integration of the semi-autonomous building blocks into a total system concept which provides for a unique operational entity. Virtually all operations contemplated for CAMDS are an improvement to and an extension of proven systems developed by the U.S. Army over the past 6 years. Numerous tests have been conducted and documented, which have resulted in the establishment of procedures governing the operations at the CAMDS at Tooele, Utah. For each and every operation planned for CAMDS, a specific series of tests has been conducted, to prove out esoteric features, safety aspects, efficiency, and human factors.

In conclusion, it is appropriate to indicate that through the years, beginning with the acceptance by the original board in 1928 of the American Table of Distances, concerted efforts have been and are being made, to provide realism and factual data in the establishment of lasting and effective chemical and explosives safety standards.

Abstract

The Department of Defense Explosives Safety Board pursuant to express statutory authority provided by 10 USC 172 and DoD Directive 5154.4 establishes uniform chemical safety standards and policies for manufacturing, testing, transportation, storage, maintenance, and demilitarization of chemical ammunition including the siting and construction requirements for chemical ammunition and explosives facilities. The standards are applicable to chemical ammunition and explosives under DoD custody throughout the world presenting real or potential hazards to life and property inside and outside of installations. A resume of the features incorporated into the chemical safety program will be presented which insures that the uses of nongovernment land around DoD ammunition installations are compatible with mission accomplishment and protection of the public. The program design assures that there are controls or minimum risks to personnel and property based upon the expected inherent toxicity or hazard of the chemical agent and explosives, the predicted concentration of agent at the event, and the expected period of potential exposure.

Literature Cited.

1. DoD 5154.4S, "DoD Ammunition and Explosives Safety Standards," March 1976.
2. DDESB Technical Paper No. 10, "Methodology for Chemical Hazard Prediction," March 1975.

RECEIVED August 2, 1978.

14

Designing a Safe Academic Chemistry Building

JOHN J. HOUSER

Department of Chemistry, The University of Akron, Akron, OH 44325

This fall the Chemistry Department at The University of Akron is leaving a building completed in 1950 at a time when the University had a student enrollment of about 3800. For the last several years, the enrollment has been approximately six times this figure. In 1964, the department commenced a doctoral program, necessitating increased research facilities. Previously, the major research effort in the department had been carried out by graduate students and faculty who were also members of the Institute of Rubber Research. Most of the research laboratories that were originally furnished with the building were designed for the rather specialized needs of this group. When the IRR became the Institute of Polymer Science in 1967 and moved into the newly constructed Science and Engineering building, the Chemistry Department took over their research laboratories and converted them in a makeshift fashion to standard chemical research laboratories. By 1974, it had become apparent that the building was wholly inadequate and outmoded, and the decision was made to replace it.

Preliminary Planning

In retrospect, it was fortunate that the Department had to wait until the 1970's for a new building because of two recent phenomena. First, the establishment of governmental regulatory agencies, most importantly OSHA, NIOSH and EPA, and their growing interest in safety and health in academic institutions as well as in industrial operations have raised the spectre of substantial fines for practices that have long been commonplace in university Chemistry departments. Secondly, there has developed over the last fifteen years or so a far greater willingness on the part of students and their parents to sue universities, faculty members and graduate assistants in cases of injury suffered in Chemistry laboratories. Recognizing that these pressures could only increase in the future, and being more aware ourselves of some of the insidious long-term effects of low levels of chemicals in the working place, the Department and the

0-8412-0481-0/79/47-096-243$05.00/0

University Administration resolved to make safety the main consideration in the design of the new facility. Accordingly, the Department's three-man Safety Committee was directed to examine several recently completed Chemistry buildings and to represent the faculty in discussions with the Administration and with the architects. It should be noted here that we were very fortunate in having as the Vice President for Planning an Organic chemist who had taught previously at another university. He consequently understood and was in complete agreement with our goals, and he effectively represented us with the rest of the Administration. He was aided in presenting our viewpoint by literature(1, 2, 3) which dealt with the problems of safe laboratory design and safe chemical handling, and with the possible consequences of a government investigation of university facilities. Of this literature, one article which can probably be singled out as having made the greatest impression is the Chemical & Engineering News Special Report, Chemical Lab Safety and the Impact of OSHA(1).

General Building Design

The University of Akron is a state university, and as such, is held to the space requirement standards of the State Board of Regents when it submits requests for funds for new buildings. These standards tie allotted space, and hence funds, to the projected enrollment ten years hence. Based on a projected enrollment increase of six-percent per year, the University in 1975 was allotted \$6.00 million for a new Chemistry building having approximately 72,000 sq. ft. Subsequently, it was decided to combine this facility with a proposed Health Sciences building, and the State approved the present 140,660 sq. ft., \$9.75 million building consisting of a four-story Chemistry wing, containing over 80,000 sq. ft., and a three-story Health Sciences wing.

The University is situated on the side of a hill, and most buildings have entrances at two levels. It was decided to take advantage of this topography to solve one of our problems. Experience had shown us that a serious difficulty with our old building was that classrooms (which are assigned to non-Chemistry as well as to Chemistry courses), undergraduate teaching laboratories, graduate research laboratories, and faculty office/laboratories were scattered throughout the building allowing people with little or no chemical knowledge to be in areas where they might be tempted to investigate the "interesting" chemical apparatus and instruments in the research laboratories. Despite attempts to keep such areas locked, minor incidents of missing chemicals and changed settings on instruments did occur from time to time. To us, the risk of damage, fire or injury seemed unacceptably high. In designing our new building, then, we decided to have all classrooms and the Introductory Chemistry laboratories restricted to the third floor. This, because of the rather steep slope of the hill at that point, happens to be only slightly lower than the first floor of one tower of the Science and Engineering building located some one hundred feet away. A covered elevated walkway, then, connects this building with the third floor of the Chemistry/Health Sciences building. At the end of this walkway, pedestrians can enter the Health Sciences wing without entering the Chemistry wing. The undergraduate Organic laboratories and the Biochemistry laboratory, to minimize large hood duct travel, are located on the fourth floor, together with two Biochemistry faculty office/laboratories, Biochemistry research laboratories and a Clean Room. Space limitations forced us to place these latter on the same floor as the undergraduate Organic laboratories, but Organic Chemistry students have had at least one year of Chemistry, and there is no reason for non-Chemistry students or casual passers-by to be on this floor. The most hazardous areas, the graduate Organic and Inorganic research laboratories, and the Chromatography/Distillation laboratory are located on the second floor together with several faculty office/laboratories. The remaining faculty office/laboratories are found on the ground floor along with the Analytical and Physical Chemistry laboratories, an NMR, ESR, UV instrument room, Chemical Stores and the main departmental office complex.

The layout of rooms and laboratories and of work areas within laboratories was intended to prevent anyone's being trapped by fire, smoke or chemical fumes. There are at least two doors to every room in which chemical work could be carried out. In cases where for security reasons a rear door is kept locked, that door is fitted with a window which in an emergency can be broken to permit access to the lock. All laboratory aisles are a minimum of five feet in width to allow rapid escape. The main entrance to all offices and laboratories is from a central hall running the length of the building and having stairwells at either end. In faculty office/laboratories and graduate research laboratories, the office areas are adjacent to the hall so that students arriving for conferences with their instructors need not enter a laboratory, and a graduate student studying in his or her office area cannot be trapped by a sudden fire in the laboratory.

Water, gas, electrical and drain lines are contained in chases which run vertically between floors on either side of the hall. Outside each laboratory there is a locked access panel in the chase, the key to which is kept in the laboratory. In an emergency any utilities to a single laboratory may be shut off from this chase. Of course, the building is equipped with the usual array of smoke and heat detectors and fire alarms. Sprinkler systems and water stand pipes were purposely omitted in the Chemistry wing to avoid accidental contact of water with active metals or metal hydrides. Cylinders supplying specialty gases or compressed air to laboratories are stored in the central hall in locked chambers located between the utility chases and identical to them in outward appearence.

Instead of the conventional ceiling-mounted, chain activated safety showers, all laboratories are furnished with one or more squeeze-operated sprayers attached to ten feet of half-inch reinforced hose anchored in the ceiling. The nozzles are of the aerating type and give a stream of water that is rapid yet gentle enough to be directed into the eyes. The feeling was that this type of emergency water supply could more easily be directed to any area of the body without soaking the person entirely, thereby reducing the tendency of the affected person to panic and resist the treatment. Should a person be extensively covered with a chemical, he or she, after preliminary rinsing with the hand sprayer, could be taken to the Clean Room which has full shower facilities and a supply of disposable clothing.

All teaching and research laboratories are provided with a color-coded "safety island," a highly visible, easily accessible area in which are located the emergency water sprayer, one or more CO_2 extinguishers, a bucket of sand and a wall-mounted fire blanket. Dry powder extinguishers were avoided because of the destructive effect of the airborne powder on electronic instruments.

A major consideration in the design of any safe Chemistry building is the ventilation and exhaust system, and it is here that a conflict between safety and energy conservation was encountered. In response to the energy crisis, the state legislature of Ohio now requires the designers of all new buildings constructed with state funds to explain in detail how the proposed structure will consume considerably less energy than comparable existing buildings. The 160-hood exhaust system envisioned for our new facility would have removed from the building enormous quantities of heated or air-conditioned air, imposing an intolerable energy burden on the University. The problem was circumvented by using induced-air hoods in all locations except for the Introductory Chemistry laboratory, the Clean Room and a few laboratories in which canopy hoods are used for temporary storage of chemicals. These hoods, manufactured by the Taylor Equipment Co. of Taylor, Texas, use 75% outside air (tempered in the winter to 60°F) and 25% room air. An automatic baffle arrangement within the hood acts to maintain this 75/25 ratio with a 100-150 cfm face velocity regardless of the position of the hood door. A manual interior baffle allows the operator to exhaust more efficiently vapors either heavier or lighter than air. The outside air is drawn in through ducts in the base of the building, filtered and blown down the outside face of the hood door. In the event that the door is completely closed, the outside air is directed down the inside face of the door, while a row of holes in the front of the hood below the door serve to maintain at least some room-to-hood air flow. The exhaust ducts from all hoods or hood groupings below the fourth floor run horizontally to the closest outside wall and into a chase running the entire height of the building. The building has eleven such chases. On the roof, each duct is connected to a separate fan. The places where hood groupings occur will be noted in the descriptions of individual laboratories. Hoods on the fourth floor are exhausted vertically to the roof. The hood system is so arranged that the hoods in each room of each floor are exhausted through separate ducts to eliminate suck-back of vapors, a not uncommon problem in buildings having several hood ducts in common. Complementing the induced-air hoods, the heating, ventilating and air-conditioning system is capable of moving air through the building at approximately three times the usual rate, ensuring that even a slightly volatile chemical spilled on a floor outside of a hood cannot build up dangerous levels of vapor over a period of time. This is particularly important where mercury is in use.

Design of Specific Areas

The paragraphs which follow will describe the particular safety features found in certain areas of the building.

Introductory Chemistry There are eight Introductory Chemistry laboratories arranged in four blocks of two, the two being separated by pocket doors. A laboratory has 24 student stations, each one with a bench-top "T" hood. The hood has an adjustable baffle to allow for removal of vapors heavier or lighter than air. Each laboratory has in addition an induced-air hood in which is stored the chemicals to be used in a given experiment. The 24 "T" hoods are exhausted as a group through a common duct; the storage hood is exhausted separately.

Organic Chemistry There are two Organic laboratories, each with 32 student stations. Following the example of at least one other Chemistry Department (The University of Pittsburgh), we have eliminated all standard benches in these laboratories, and the students do all experiments in eight-foot induced-air hoods, two students to a hood. The hoods are exhausted in groups of two. There are several advantages to this approach. First, barring the occasional dropped flask, the traditional overpowering odor and toxicity hazard of the undergraduate Organic laboratory should be essentially eliminated. Spillage on the laboratory floor is reduced since the hood floors have an approximately one-inch lip. The danger of fire is greatly lessened, both because burners will be replaced by heating mantles, and because solvent vapors cannot accumulate. The hood doors, when pulled down, provide each student with a safety shield. Should a fire break out, it is easily confined to a small area and readily extinguished. Finally, to most undergraduate students, a hood is a place to store malodorous or hazardous chemicals. It is hoped that forcing students to work in hoods will accustom them to the advantages thereof and lead them in any subsequent laboratory work to look for a suitable hood before beginning even to collect the necessary chemicals.

A six-foot canopy hood mounted over a bench is situated against one wall. This bench is used to store general solvents and any chemicals needed for a particular experiment. A custom-made seven-foot sloped stone drain board and sink is used to store aqueous inorganic acids, bases, etc. An ordinary kitchen sink sprayer is provided for flushing spills away. At either end of each laboratory are tables for melting point apparatuses, rotary evaporators and other small equipment. The evaporators are intended to overcome the student's natural tendency to boil off solvents into the air--a practice considered undesirable even in a hood. The two Organic laboratories are separated by an instrument room which houses an NMR, two IR spectrometers and two gas chromatographs.

Analytical Chemistry The Analytical Chemistry laboratory with 26 student stations contains, along one wall, two six-foot bench-mounted induced-air hoods and two six-foot walk-in hoods. These hoods are exhausted in pairs.

Physical Chemistry/Instrumental Analysis The Physical Chemistry laboratory, containing 20 student stations, is used also for Instrumental Analysis and for certain courses in the Community and Technical College which administers two-year associate degree programs. The laboratory is equipped with three four-foot induced-air hoods and a five- and a six-foot walk-in hood, each with a built-in vacuum rack. Against one wall are located four peninsular work benches ventilated by a single canopy hood. Against another wall, a microscopy table and a refractory table are also ventilated by a common canopy hood. All canopy hoods are exhausted together.

Biochemistry The Biochemistry laboratory has 30 student stations and also serves the Community and Technical program as well as undergraduate Biochemistry students. Four six-foot and one four-foot bench-top induced-air hoods are located against one wall. Three of the six-foot hoods are exhausted together.

Graduate Research With one exception, the graduate research laboratories are 24' x 30' and are designed for occupancy by four students. It was felt that having a series of two-student laboratories would result in a needless duplication of equipment, while placing more than four people in a single room would lead to overcrowding. Our experience with hood usage in the old Chemistry building was that most experiments were carried out on bench tops because the hoods were used mostly for chemical storage. This practice had the further disadvantage that it required the hoods to be left on continuously. Our solution to this in the new building was to place five four-foot hoods in each four-student laboratory. Four of the hoods, exhausted together, are used for conducting experiments and have the entire complement of utilities. They are turned on only when actually in use. The fifth hood contains no utilities, but rather is fitted with shelves and is used only for storage of chemicals. This hood is exhausted separately and is left on continuously. Each research laboratory contains in addition a 3' x 7' x 1½' safety cabinet for storage of solvents. This airtight steel cabinet meets OSHA standards and is designed to suffocate a fire, should one start within it, and to contain spills from leaking bottles.

Faculty Office/Laboratories The 12' x 30' faculty office/ laboratories are divided into a 12' x 10' office separated from a 12' x 20' laboratory by a door. The rear door of the laboratory section typically opens into an adjacent graduate research laboratory. The faculty member is provided with a four-foot induced-air hood, but is expected to use the chemical storage hood and solvent safety cabinet in the graduate laboratory.

Clean Room The Clean Room was included in the building in response to the published OSHA regulations for the handling of carcinogenic and otherwise highly toxic materials in industrial laboratories. Our expectation is that these regulations will in the future be extended to academic institutions. The walls of this 12' x 30' room are coved floor and ceiling and are covered with sheet vinyl as is the floor. The suspended ceiling uses replaceable gypsum board panels. For ease of decontamination, the laboratory work bench and sink are constructed of stainless steel as are a glove box and a hood. This radiological-grade hood is not of the induced-air type and is not even exhausted to the outside. Rather it is exhausted back into the room through a series of replaceable filters. In the front of this room is a tiled shower and dressing room. The Clean Room is located directly across the hall from the undergraduate Organic laboratories, making it very convenient for any student who has been badly splashed with a chemical to shower immediately and get a change of clothing. The privacy of this room should eliminate any reluctance on the part of the student to wash thoroughly enough to avoid serious injury. It must be emphasized that the Clean Room is for use with toxic materials only, not explosive ones. A "High Pressure" laboratory having steel walls is available in the Science and Engineering building for this purpose.

Chemical Stores The Chemistry Department has the responsibility of providing chemicals and chemical equipment to the entire campus. This necessitates storage and handling of much larger amounts of these items than would normally be required for a department of this size. In the new building, the Chemical Stores operation is contained in a one-story wing located in the middle of the first floor near the elevator and adjacent to a loading dock. The wing is landscaped on the three exposed sides with an earthen bank which extends to within three feet of the eleven-foot roof line. The roof itself is equipped with ventilation fans and pressure-releasing plastic blow-out panels. Inside, the area is divided into a receiving room, a waste-solvent holding room, and three separate storage rooms for dry chemicals, liquid chemicals and equipment, including glassware. Within the chemical storage rooms, materials are grouped generally alphabetically but with overriding consideration given to compatibility. Should a fire break out in any of the storage areas, the smoke and heat detectors will trigger a CO_2 extinguisher system located in the ceiling and supplied from cylinders stored in the receiving room. At the same time, an annunciator system located in the Chemical Stores office will sound an alarm, and by means of a lighted panel will show the location of the fire. To reduce still further the fire danger, all phones, switches, receptables and light fixtures in the Chemical Stores area are static free. The waste solvent stored in the holding room is contained in five-gallon polyethylene cans and is picked up weekly by a private chemical disposal service. These safety solvent cans with flame arresting openings and spring-loaded caps are placed in every laboratory in the building. Enough extra safety cans were purchased so that a full can may be replaced immediately. Waste solid chemicals in properly labeled containers, are also picked up for disposal in a landfill.

Summary and Conclusion

Recognizing that external pressures are forcing academic Chemistry departments to abandon their traditional relaxed attitudes toward safety, we have attempted to construct the safest building possible with the available funds. This alone will not guarantee accident- and hazard-free operation, however. The faculty, administration and staff must support and encourage safe practices. In this spirit, the University in 1977 established a campus-wide safety committee to assist and advise departments and the various support services when questions on safety arise. In addition, money for safety equipment is now furnished from a special fund established by the Board of Trustees and is no longer charged against departmental operating budgets. At the educational level, for some years now, all incoming graduate students and stockroom personnel in the Chemistry Department have been given a safety orientation in which the whys and hows of safe practices are discussed. Any particular hazards of an undergraduate experiment are pointed out to the students beforehand, and as previously unrecognized problems with common chemicals become known, experiments are modified or dropped. For example, benzene and chloroform have been replaced as solvents in the undergraduate Organic laboratory with toluene and methylene chloride, respectively. Accident reports are filled out whenever mishaps do occur, and these reports are reviewed periodically in an attempt to identify problem areas.

Finally, the Department of Chemistry is pleased to share its experiences and ideas with others and to answer any questions about the design and construction of our new building. Answers to technical questions concerning construction details can be obtained from the architectural firm, Tuchman, Canute and Wyatt of Akron, Ohio, or the consulting engineers, Schiesser, Buckley and Keyser, also of Akron. Visits from our colleagues are invited and encouraged.

Literature Cited

1 Anonymous, C&EN, May 24, 1976.

2 U.S. Department of Labor, Occupational Safety and Health Administration, General Industry Safety & Health Regulations, Part 1910, June, 1974, OSHA 2206.

3 American Chemical Society Committee on Chemical Safety, Safety in Academic Chemistry Laboratories, March, 1974, revised, January, 1976.

RECEIVED December 11, 1978.

Flexible Laboratory Design within the Constraints of Safety

PAUL D. GARN

The University of Akron, Akron, OH 44325

Remaining a free spirit in a more and more regimented society isn't really possible but some freedom and efficiency of operation can still be maintained. My work is principally with instruments but being a chemist I am obviously a very hazardous person; hence I must be protected against myself just as much as a preparative organic chemist or a toxicologist. To clarify the image, I should point out that I am certainly pro-safety. I installed -- and used -- seat belts in my car about 1960 and the buzzer in my 1973 model has not been disconnected. I have had several cuts and burns in the laboratory but never a lost time accident -- nor has anyone under my supervision. But the design features set up for ordinary chemical operations are sometimes a needless expense for my kind of research and certainly a laboratory set up for organic research would contain many impediments to instrumental use and design.

My objects in presenting this material are to remind designers, architects and planners that a stereotyped laboratory does not provide for effective use of space and to suggest to users that there are cost-effective ways to elude the more-confining designs and still work within the system.

Twenty-six years of bumping around in cluttered industrial and academic laboratories has given me a considerable impatience with "standard" laboratory designs. Early in my career at Bell Labs, I learned the value of elbow room and non-fixed positioning because I had neither. Instrument design being an important phase of my work then as well as now, the ability to open the back or attach something to the side of an apparatus is very important. When I had an opportunity to lay out a laboratory, I tried to eliminate a service island in favor of strategically placed services and to

0-8412-0481-0/79/47-096-253$05.00/0

get a four-bay laboratory instead of a three or a five. I learned immediately that our space patrol was adamant about deviations from standard arrangements. With service island firmly implanted in a three-bay lab, I gained some mobility by moving out a couple of benches but when I began to use a variety of gases the clutter came back.

About 1958, I was given the task of designing a fluorine laboratory wherein reactions would be conducted in atmospheres of fluorine, chlorine trifluoride and/or hydrogen fluoride. Expecting this to be a closely supervised operation, I and a member of the planning staff laid out a system of controls on one end of a twelve foot walk-in hood. All the working cylinders were housed in a closet accessible only from the hall and separately ventilated. That got rid of the major clutter, there being several access ports along the back wall through which any chosen gas or mixture could be directed to any of the reaction vessels, two were constructed and space was left for others.

Actually I never operated the laboratory -- when I learned that the level of supervision was to be less than I had anticipated I got out of the fluorine business. I was next door for my remaining three years at Bell and there were no accidents in that time.

Not long after coming to The University of Akron, I was in the design business again. The department was expanding but the building was not. However, we had several classrooms and some new buildings were providing more, so I had the task of converting one to a laboratory suitable for instruments. Still thinking in terms of a quasi-stationary state of research use, I drew up general plans involving gas facilities being supplied from closets outside, one being reserved for hydrogen and other flammables. The closets are shown in Figure 1.

The gas facilities comprised eight lines of hard copper tubing, each appearing at two locations in the laboratory. Some of these have had the same gases in them for several years. It is possible that one has not been used at all. For others, though, the use has varied -- not from month to month very often, but perhaps a few times a year.

I had these, electricity, exhaust, and water and drain, being supplied overhead, so that instruments could be placed wherever needed. See Figure 2.

I had specified a number of benches and given a tentative arrangement. The outline of needs then went to the architects who specified fittings and drew the

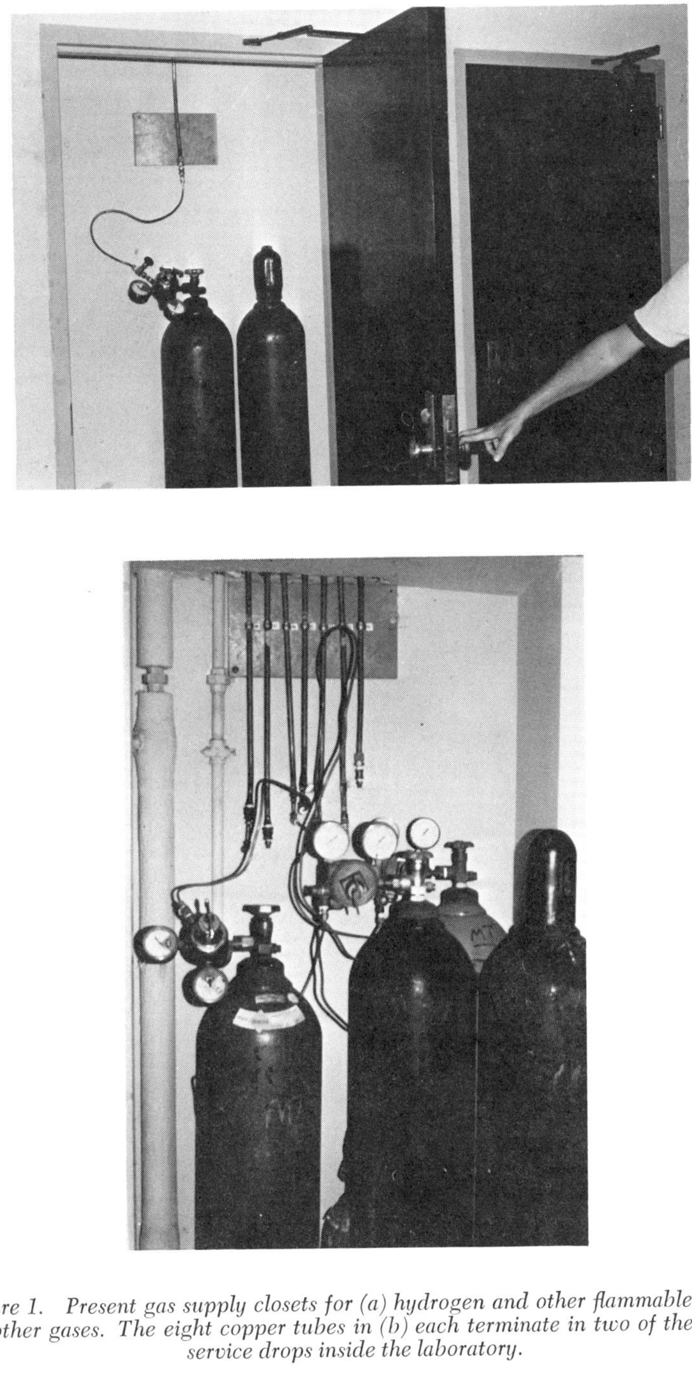

Figure 1. Present gas supply closets for (a) hydrogen and other flammables, and (b) other gases. The eight copper tubes in (b) each terminate in two of the eight service drops inside the laboratory.

Figure 2. Service drops in instrumental laboratory. (a) Gas and electrical supply for differential thermal analysis and (b) cooling water and electrical connections to a mass spectrometer.

final drawings and oversaw the actual construction. When the contract was completed, I was let back in.

The service drops were nicely within reach and the valves were pointing downward. I could walk under one so long as the first fitting was an elbow or the tubing was bent. The benches were not simply placed in islands; they were screwed to the floor. Not only that -- where two or three or four were together they had a single top. I'm just building up to a climax.

In spite of these lesser annoyances, I moved in some equipment and began hooking it up. To supply helium to a gas chromatograph, I connected a cylinder via copper tubing to the supply line, turned on the cylinder, set the pressure at fifty psi, went in to the lab to get the air out of the lines, turned the valve -- and took a shower without soap. The standard leak test is -- or was -- with water. No one had suggested to the contractor that the stuff ought to be dumped.

Several days and cylinders of nitrogen later I was able to begin using the laboratory. In spite of the errors, the experience with the ceiling-mounted facilities has been quite generally positive.

During these last fifteen years I've become aware of a number of differences between industrial and academic laboratories; unless a person is very well funded the research goes in fits and starts. There is little continuity as students come and go. Each student is likely to want to work on something different so flexibility is mandatory and ease of change or beginning is certainly helpful. Occasionally a new idea takes form. For these reasons I want to maintain the freedom of movement within the laboratory. Partly with this in mind I chose not to have new lab benches in the new laboratory. Instead, I will have very few permanent installations, then I will bring over the benches <u>after</u> the contractors leave. We will dump most of the mementos left by past graduate students somewhere along the way.

My labs will have just one bench hood and one walk-in hood (Figure 3) but a number of places to connect flexible exhausts (Figure 4). The building is just being completed at the time of writing so the description must be of plans as compared to the finished product we expected to occupy during the summer.

One student is preparing liquid crystals having multiple smetic states in order to study pressure effects on the smetic-smetic transitions but that is only organic preparative work in my group in years. Another is preparing and studying inorganic complexes.

For other work we occasionally use organic

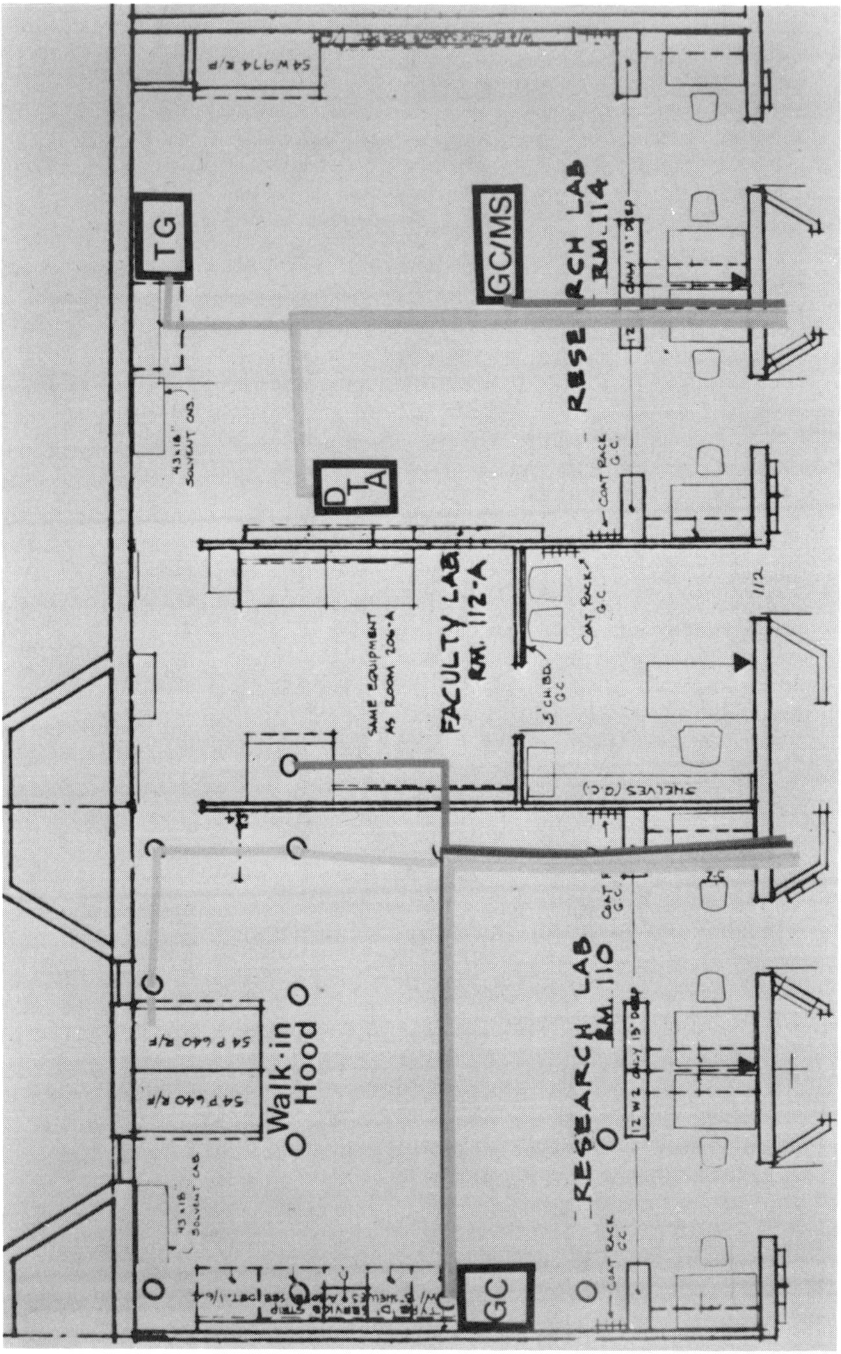

Figure 3. General plan of office and research laboratories showing the hoods and paths of gas lines

Figure 4. View of overhead facilities showing exhaust and electrical service channel

Figure 5. "Exterior view" gained by use of an 8 × 12 foot mural—temporarily mounted for this photograph

solvents within apparatus but the flexible exhausts will take care of these uses.

One use here will be for a controlled atmosphere thermobalance I am building. Some of the atmospheres I plan to use will be quite noxious. I'll need exhaust some of the time but I do not want the thermobalance in a hood.

Another study, electrochemical this time, introduces the possibility of generating chlorine gas if the microprocessor control apparatus fails or if the software is not adequate. Not yet being proficient in such programming, I am designing for a mobile system I can put in the walk-in hood. A mobile apparatus is also planned for some work using carbon monoxide. Alarms in closet, lab, and hood should give adequate notice of any escape. In the present laboratory, the exhaust stream has been through copper tubing thrust almost a meter into one of the flexible exhausts.

The gas services will again come from closets but this time each cylinder-to-apparatus connection will be made with tubing -- made by the student -- or me -- when it is needed. For neatness the tubing will be supported by hangers dropped down from the ceiling.

I plan to go back to colored plastic tubing for those uses for which it is adequate. Where the tubing must pass over or through the graduate student's offices I plan for it to go through plastic drain pipe. Where it must pass over a partition a couple of coupled elbows will guide it.

I have already found one result of inadequate communication between user and designer. The electrical service channel is not just overhead -- it is at ceiling height. I will have to decide whether to recruit graduate students from basketball teams, buy a ladder if there is money left, or have the channels lowered.

A similar plan will be used for gases in the analytical teaching laboratory. There are occasional needs for gases for chromatographs and such but several classes use the lab. We already use instruments on carts for the chemistry majors to avoid taking up lab space during other classes. Hooking up one or two gas lines takes only a few minutes. In a few cases, I will run lines to valves secured to the service strip to avoid the need to flush out long lines.

Finally, the prospect of an inside office failed to intrigue me, but neither did the idea of looking at the wall of the adjacent building. At Bell Labs one summer, I could watch a hummingbird care for her young. Since I couldn't go to the mountain I am bringing the mountain to me by putting up a giant mural with a few meters of material to simulate drapes (Figure 5). I toyed with the idea of draw drapes until I checked prices. This is our chemical stores manager lifting up his eyes unto the hills -- which were taped to the walls for the photograph.

To Summarize:

Laboratory design needs to take account of varied uses even at the initial stages. For example, to move into an already fitted lab would require removal of some facilities -- an additional cost.

Instrument rooms need floor space more than service islands. The services can as well be brought down individually from overhead. Flexibility of use is easily achieved if there are not too many permanent fixtures in the way. To move into an already fitted lab would have required removal of some facilities -- an additional and unnecessary cost.

With the present arrangement, I can interchange the several furnace assemblies I use with the differential thermal analysis; I can move a data acquisition system from apparatus to apparatus; I can put reaction apparatus in position for direct monitoring by mass spectrometry; and I can use the exhaust facilities effectively. In addition, I saved money which can be put to better use.

RECEIVED November 22, 1978.

16

Handling and Transport of Hazardous Materials

HOWARD H. FAWCETT
Equitable Environmental Health, 6000 Executive Blvd., Rockville, MD 20852

WILLIAM S. WOOD
1076 Dunvegan West, West Chester, PA 19380

Abstract

The design of loading and unloading facilities is critical to the safe transportation of hazardous chemical cargo. Adequate loading racks and transfer equipment, designed to protect employees and the cargo during pumping operations, are being built by some companies. Advances in ullage control, fire protection, respiratory protective equipment, safety showers and other equipment and facilities contribute to personnel safety, property conservation and economic efficiency in operations.

Tank vehicle cleaning and inspection operations frequently can be improved by better methods and supervision. Handling and warehousing of drummed and packaged goods have been upgraded by modern methods, but education, supervision and control of the human element is still a substantial challenge.

Handling and Transport of Hazardous Materials

Transportation of hazardous materials is a vital function of our economy and our way of life. Most shipments reach their ultimate destinations without incident, but the few that derail, spill, explode or drive people from their homes have become frequent news items, especially in the past few years.

(The opinions expressed by the authors are based on personal observations and do not necessarily reflect the Organizations' with which they are affiliated.)

0-8412-0481-0/79/47-096-263$05.00/0

However, the operations of loading, unloading, storing and sampling have more accident potential than the movement of cargo in closed tanks once they are loaded.

Tank car and tank truck loading and unloading, gaging and sampling of liquid cargo inherently requires that personnel work on top of the tank cars or trucks. There are numerous cases of falls from top dome locations due sometimes to exposure to the material being transferred, or to the weather conditions, such as rain or ice. A few tank cars are fitted with platforms surrounding the dome. A well designed fixed rack with a swing platform and guardrail is highly desirable, but often not available. The Manufacturing Chemists Association has published guidelines for such facilities (1). Similar racks with platforms are needed for tank trucks.

Swing lines made up of pipe with flexible joints and provided with a counterbalance is the preferred method of transferring hazardous chemicals into or from a tank vehicle. Hoses, no matter how thoroughly reinforced, have lower pressure capability and are easily damaged in use. Rupture of transfer hose is a not uncommon cause of employee injury or spill of product. Metallic tubing with braided exterior has admirable qualities, but a sharp bend, particularly near the end, causes weakening and failure, which is often neglected before the failure.

Environmental standards prohibit spillage that can reach a waterway; hence, loading and unloading areas must be paved or otherwise secured and drained to recover or control any losses. Flushing of the site with water to wash off spilled material may tax recovery capacity or complicate subsequent treatment and disposal. Tank truck terminals have a real problem in treatment and disposal of drainings and washings derived from tanker cleaning operations, as do tank cars (rail) and tanker-barge operations (waterways). (2)

A special accident potential exists in the manual cleaning of tank vehicles of all types and sizes, due to the use of gases other than air used for pressure unloading. Liquid cargo, whether hazardous or not, is often unloaded by applying nitrogen or other "inert" gas to the tank in order to displace or push out the contents. An unsuspecting tank cleaner who enters such a vessel to clean it is likely to suffocate and die without knowing he was in trouble. In many recorded incidents, his would-be rescuer, entering without respiratory protection, such as self-contained breathing appartus, or assistance, meets a similar fate. The hazards of entering any enclosed space are recognized in most plants and proper precautions are taken (3), but these fatalities continue to occur.

How does the pumper know when the tank vessel is full? Tank cars and trucks for volatile liquids may have permanently attached fittings and no provision for opening the dome.

If any means of gaging is provided, it is a dip tube that blows off vapor until the liquid level reaches it. The possibility of error is significant. Overfilling frequently occurs.

Two feasible methods of supplementary gaging are suggested. Load cells (strain gages) placed under the truck or under the truck wheels can be used to indicate the weight of contents continuously and to sound an alarm when the desired weight has been loaded. Another arrangement consists of a radiation source and a geiger counter mounted on an arm that could be lowered across the top of a vessel being loaded. When liquid reaches the height of the source and counter, its mass will attenuate the radiation and a signal can be displayed or made audible. The radioactive source, usually an isotope, requires licensing and may not be suitable for some users.

Locations where flammable liquids are loaded or unloaded should have adequate fire protection. A water spray system operated manually and automatically is the most economical and best for most materials. Where water supply is inadequate or unavailable, a dry chemical system can be engineered. Ultraviolet ray detectors are recommended for speed of operation and automatic discharge of the agent.

Minimal personal protection for loading racks will include safety shower or showers, eye wash facilities, and emergency respiratory protection. At least two self-contained breathing apparatus, preferably of the pressure-demand type, should be available at the site, and employees should be thoroughly trained in its use. Where corrosive materials or chemicals toxic by skin absorption are being transferred, a full suit of protective clothing should be available. Care must be taken to insure that the protective clothing, such as gloves, actually is impervious to the substance to be handled; many elastomers in use are not impervious to certain substances. In northern climates, safety showers and eye-wash equipment must be protected from freezing. One approach is to use low voltage induction heating on all exposed piping. In some isolated locations, no water is available, and provision of a safety shower requires use of a tank of pressurized water, preferably tempered for comfort and to prevent freezing.

In the petroleum industry, unattended bulk stations exist where the tank truck driver is alone when he selects cargo, connects lines, starts pumping, gages contents, stops pump, disconnects, and drives away. If an error or accident occurs, he is on his own. Some type of alarm or closed circuit television monitored at a nearby office or communications center would be invaluable in case of an emergency.

Placard and label requirements for interstate shipment were changed by the U. S. Department of Transportation effective January 3, 1977 (3).

One of the objectives was to accomodate to United Nations labeling for international shipping using symbolic representation of hazard. The new labels and placards overcome any language barriers that may exist.

Drum and package shipment of hazardous materials involve a number of hazard potentials. For example:

> What cargo items can be safely shipped in the same car or truck? (2)
>
> Can salamanders used to prevent freezing be used when hazardous chemicals are on board?
>
> What limitations apply to radioactive cargo?
>
> How must heavy items such as gas cylinders be secured in the truck or car?
>
> How is the van placarded when mixed cargo such as flammable, toxic, reactive, etc., are shipped together? (2a)
>
> How should materials be segregated in terminals?
>
> Are ordinary fork trucks suitable for drums of flammable liquids?
>
> Do terminals have suitable fire protection for flammable liquids, or ventilation suitable for toxic volatiles?
>
> What quantities of chemicals may be shipped by the various modes, (i.e., truck, barge, rail, passenger aircraft, cargo aircraft)?

Consider the plight of shipping clerk, or dock worker or truck driver who may be called upon to make some of the decisions indicated. With limited education, sometimes inadequate information, voluminous regulations, and under time pressure, mistakes occasionally occur.

Fortunately, DOT requires that persons responsible for such shipping operations be trained and a number of training programs have become available. One of the most recent is a seminar presented by the Operations Council of the American Trucking Associations, Inc. (4) (4a)

Much discussion has evolved about the regrettable condition of the railbeds of our country. Derailments occur daily. Much of the rolling stock is also seriously in need of repair. Essentially all tank cars are owned or leased by shippers; hence, can be improved at industry expense. Boxcars, gondolas, etc., are usually owned by the railroads (or Conrail) and a broken axle, hotbox or other failure of <u>any</u> car can cause a pile-up, possibly involving tank equipment or trackside hazards.

In a derailment there is a tendency for couplers to move vertically and come apart. The free coupler then becomes a battering ram that can easily breach a tank car. Shelf couplers prevent separation, and end shields on tank cars can prevent penetration and spill, and are now required by DOT under recommendation of the National Transportation Safety Board. (5) Such improvements were developed by the Association of American Railroads in 1970, but implementation has not been pushed.

Flammable liquid spilled in a wreck sometimes causes overheating and failure (or BLEVE) of tank cars. (6) Insulation to retard heat transfer gives emergency crews time to effect control before overheating failure. All three of these measures--shelf couplers, end shields and insulation--will soon be required on tank cars.

Our highway system is not quite perfect, either. Have you ever negotiated a cloverleaf in your car at the recommended speed and wondered if you were going to make it? Consider then a truck with high center of gravity and maybe a sloshing partial load. High crown secondary roads and narrow bridges are only a couple of more accident potentials faced by the drivers. Speed limits are not always observed, and trucks overtaking a passenger car create additional hazards, especially in inclement weather.

Police and firemen are called whenever an accident or spill occurs in the transportation network. They desperately need identification and specific on-scene coordination of activities involving cargoes and materials (7), of involved cargo and hazard properties of materials.

Eight years ago, the Manufacturing Chemists Association organized CHEMTREC (Chemical Transportation Emergency Center) on the recommendation of the NAS-NRC Committee on Hazardous Materials to respond to calls from emergency personnel with information regarding hazards and recommendation for controlling the emergency and evacuating residents. The toll free number is <u>800/424-9300</u>. Over 3,500 chemicals are immediately "on call", with a referral to the manufacturer if additional help is needed.

In addition to CHEMTREC, other emergency aids are finding wide use. The U. S. Coast Guard CHRIS (Chemical Hazard Response Information System) is available through the Coast Guard Headquarters, and regional officers, and its published material is of value in water-borne spill responses. (8)

The EPA-OHM-TADS (Oil and Hazardous Materials Technical Assistance Data System) is an on-line computerized system containing 126 parameters for over 1,000 chemicals, and includes recommendations for emergency control, amelioration, clean-up and disposal. It is also a useful data source for nonemergency use. (9)

Over 500 Poison Control Centers operate in the U. S. and Canada; these centers provide specific recommendations on emergency measures for physicians and other medical personnel. In general, they are located in the larger hospitals and coordinated by the State Health Departments. (10)

The U. S. Army Technical Escort Center, at Edgewood Arsenal, will dispatch assistance if requested by the Department of Transportation or by the Governor of a state.

Quite recently the National Fire Protection Association developed, under a DOT contract, a self-study course for firemen on responding to transportation emergencies. The fire services should be encouraged to include this and related material into their training. (11)

The Department of Transportation publishes "Emergency Action Guide for Selected Hazardous Materials", which describes hazards, recommended action and evacuation diagrams for 42 common bulk cargoes. This booklet should be in every fire chief's car, and deserves a very wide distribution. (12)

Tank truck drivers have been observed loading or discharging their cargoes occasionally with inadequate training or instruction. A true story illustrates what can happen:

> A driver having worked since 7 A.M. was assigned to pick up 11 tons of 93% sulfuric acid late in the afternoon and deliver it to a plant thirty miles away. The driver had no training with sulfuric acid except that he had served as helper on one previous delivery. His helper had no experience with this cargo. The pickup and transport went smoothly and the men discharged the acid into the customer's tank by pressurizing the trailer with a plant air line. At about 8 o'clock (13-hour day) the men disconnected the lines without releasing all pressure from their trailer. Both men were sprayed with acid and, since their protective suits and face shields were neatly

> stored in the truck cab, both received severe burns. They ran past the safety shower 15 feet from their truck to get help at a boiler room a hundred feet distant. The men were tired. They had not been trained in use of protective clothing. They did not know proper unloading procedures and they did not know the simple first aid for acid splash. (13)

In summary, insofar as handling and transportation are considered:

> Excellent designs for loading racks are readily available--they need more effective application. Environmental needs must be satisfied.
>
> Many freight terminals need upgrading for handling of hazardous packaged goods.
>
> Tank cars and tank trucks are being improved to preserve integrity in an accident.
>
> Rail lines and highways need more improvement to improve their utility. Barge and ship terminals and bulk-storage facilities should be continually reevaluated from the safety and efficiency aspect, since even larger potential for disaster accompany larger amounts of hazardous materials.
>
> Courses are available to train warehouse and shipping personnel regarding compliance with hazardous materials regulations. (11)
>
> Rail, trucking and other transportation employees need more than minimal on-the-job training for handling hazardous cargo and dealing with emergencies. Retraining and reorientation is essential to update instructions and to increase safety awareness.
>
> Emergency response personnel now have access to excellent training for controlling transportation accidents.
>
> Hardware improvements being developed, but application is often not rapid. Sofeware improvements are available now through training.

Chemical Incompatability

When two or more chemicals are mixed or allowed to come into contact, the possibility of chemical reaction generating gases, fumes, fires or explosions must always be considered. (14)

In bulk transportation, where large quantities are handled, and the possible loading or unloading of a cargo into a tank which may have or still contains another product, is extremely serious. To cite a few examples from the recent past:

A tank ship loaded a cargo of liquid sulfur (temperature 260°- 280°F) (127°- 138°C) on to a heel of crude oil. The reaction between liquid sulfur and petroleum generates toxic gases, including hydrogen sulfide, carbon disulfide and sulfur dioxide. The ship's crew was hospitalized and the ship, seeking a port of refuge, was denied entrance to several harbors until the reaction had run to completion as the sulfur cooled and eventually solidified.

On three separate incidents since 1971, tank truck drivers attempted to discharge their cargo of sodium hydrosulfite to a manufacturing plant. The unloading port into the building was next to an unloading port for acid. No marking or other identifications were obvious on the two connections. The driver attached the hydrosulfide to the acid connection. In each case, the generation of hydrogen sulfide caused the evacuation of the facility, but, unfortunately, not in time to prevent fatal exposure and serious injuries.

A similar incident occurred some years ago at an electronics plant in Brooklyn; the chemicals were nitric acid being pumped through an unmarked pipe into an ethanol storage tank.

Perhaps the most tragic uncontrolled reaction, and certainly potentially one of the most serious for society, occurred in 1972 as a result of improper packing of chemicals to be shipped by air. A shipper in Los Angeles had assembled a mixed cargo of chemicals for shipment to an electronics plant in Germany. The shipment went by cargo aircraft to New York (JFK), where it was transferred to a Pan Am Boeing 707 cargo plane. The three crew members took off for Europe, but over Connecticut, noticed brown fumes in the cargo compartment. By the time Boston was reached, the fumes and, by then fire, had become so great that an emergency crash landing was made at Boston Logan Airport. The three crew members were either killed (or possibly dead before the crash); the aircraft was destroyed. Investigation disclosed that over 1,000 pounds of reagent grade nitric acid had been packed in wooden boxes, using sawdust as the filler around the glass bottles. The breakage of one or more bottles apparently triggered the reactions that caused the plane to be filled with brown oxides of nitrogen and other vapors from plastic bottles of solvents; many unlabeled.

Another common chemical which has caused serious problems in handling and transport is calcium hypochlorite. When dry and uncontaminated, the chemical is safe, but even small amounts of contamination on a plastic scoop or from a soft drink bottle has been known to cause ignition. An extensive literature search of this chemical's instability is available. (15)

The aforementioned examples will serve to illustrate that chemistry--even elementary chemistry--has an important role in the transport, handling and emergency control of hazardous materials, and that the traffic, shipping and distribution departments of even the larger companies will benefit both in decreased losses due to accidents and from protection against penalties due to spills, if more attention is given to properly and continuously training personnel at all levels. Such training, whether in-house or on the job, or at seminars and conferences, is a critically needed component in our complex world. There is no lack of information resources--but the day-by-day utilization of that data base needs increased attention. (16)

References

1. Manual Sheet TC-7, "Tank Car Approach Platforms", Manufacturing Chemists Association, 1825 Connecticut Ave., N.W., Washington, DC 20009 (1974).

2. 49CFR, Parts 171-177, "Hazardous Materials Regulations", Federal Register; Vol. 41, No. 188; Sept. 27, 1976; pp. 42364 to 42638. Federal Register; Vol. 43, No. 49; March 13, 1978; Part II; Environmental Protection Agency; pp. 10474 to 10508 ("Penalties for Accidental Discharge Into Waterways"). See also: "Proceedings of the 1978 National Conference on Control of Hazardous Material Spills", Miami Beach; April 11 - 13, 1978; available from Information Transfer, Inc., 1160 Rockville Pike, Rockville, MD 20852.

2a. An excellent wall placard which describes the labels; pictures them; and, suggests basic emergency action is available from JODY, Inc., P. O. Box 88884, Atlanta, GA 30338.

3. Safety Guide SG-10, "Entering Tanks and Other Enclosed Spaces", Manufacturing Chemists Association. (1961) (Under revision.)

4. Contact National Tank Truck Carriers Assoc., c/o American Trucking Assoc., 1616 "P" Street, N.W., Washington, DC 20036.

4a. "18 Wheelers Create Instant Inferno", by Peige, J. D., Maryland Fire and Rescue Institute Bulletin; Vol. 7, No. 8; Aug. 1978; College Park, MD 20742.

5. National Transportation Safety Board Report No. NTSB, SEE 78-2; June 23, 1978.

6. BLEVE - boiling liquid expansion vapor explosion. The National Fire Protection Association, 470 Atlantic Ave., Boston, MA 02210, has comprehensive literature and training aids on this subject and related problems in handling of hazardous cargoes.

7. National Transportation Safety Board, Safety Recommendation I-78; pp. 14 to 16; August 30, 1978.

8. "CHRIS Response Methods Handbook", Department of Transportation, Coast Guard, CG-446-4; January 1975. See also: CG-446-1, a condensed guide to Chemical Hazards; CG-446-2, "Hazardous Chemical Data", and, CG-446-3, "Hazard Assessment Handbook".

9. OHM-TADS System is available to subscribers who finish their own terminal software. For details on OHM-TADS and availability, contact Informatics, Inc., 6011 Executive Blvd., Rockville, MD 20852, attn: E. Pascal.

10. "Directory of Poison Control Centers", August 1978, National Clearinghouse for Poison Control Centers, U. S. Department of of H.E.W., Food and Drug Admin., 5401 Westbard Ave., Bethesda, MD 20016.

11. Contact the National Fire Protection Assoc., 470 Atlantic Ave., Boston, MA 02210, for details.

12. Available from U. S. Department of Transportation, National Highway Traffic Safety Administration and Materials Transportation Bureau, 400 Seventh Street, S.W., Washington, D.C. 20590.

13. See the Manufacturing Chemists Association's "Safety Data Sheet on Sulphuric Acid", SD-20, and other chemicals, 1825 Connecticut Ave., N.W., Washington, DC 20009. See also, National Safety Council's Chemical Data Sheets, including, "Opening Pipelines and Connected Equipment", available from Chemical Section, National Safety Council, 444 N. Michigan Ave., Chicago, IL 60611; and, Chemical Safety References, Data Sheet 486 (Revised).

14. Flynn, J. P. and Morrisette, M. D., "Development of a Compatibility Guide for the Water Transport of Bulk Chemical Cargoes", Journal of Hazardous Materials, Vol. 1, No. 4. See also, NFPA No. 491-M, "Hazardous Chemical Reactions", National Fire Protection Assoc., 470 Atlantic Ave., Boston, MA 02210.

15. Clancy, V. J., "Fire Hazards of Calcium Hypochlorite", Journal of Hazardous Materials, Vol. 1, No. 1, pp. 83 to 94, (Sep. 1975).

16. Fawcett, H. H. and Wood, W. S., "Safety and Accident Prevention in Chemical Operations", Wiley-Interscience, 1965. (New edition is in preparation.)

RECEIVED December 1, 1978.

17

Concept Design Criteria for Standard Chemical Maintenance Facility

JAMES P. HENDRICKSON

DARCOM Ammunition Center, SARAC-DEV, Savanna, IL 61074

Toxic chemical munitions, as well as other munitions within the stockpile, require periodic surveillance together with care and preservation, to be maintained in a serviceable and issuable condition.

TASKING:

With limited facilities currently available for chemical maintenance, the DARCOM Ammunition Center, Savanna, IL, at the direction of the US Army Materiel Development and Readiness Command, was tasked to develop a concept design for a Standard Maintenance Facility for Toxic Chemical Munitions.

The transportation restrictions prohibiting the movement of toxic chemical munitions together with establishing that most maintenance and surveillance requirements are typically the same at storage installations, resulted in developing a standard modular facility design which could be sited at more than one installation.

FACILITY OBJECTIVES:

The primary objective and approach in developing a total functional facility was (Figure 1) to provide a safe and secure environment in which maintenance, preservation, and packaging, together with surveillance and assembly of chemical munitions, could be performed in an efficient and productive manner.

Facility design criteria was developed based on a variety of considerations, that would both satisfy operational and processing requirements, and comply with safety, security, and surety regulatory guidance.

It is important to note that demilitarization of chemical munitions was not considered as a requirement for this facility.

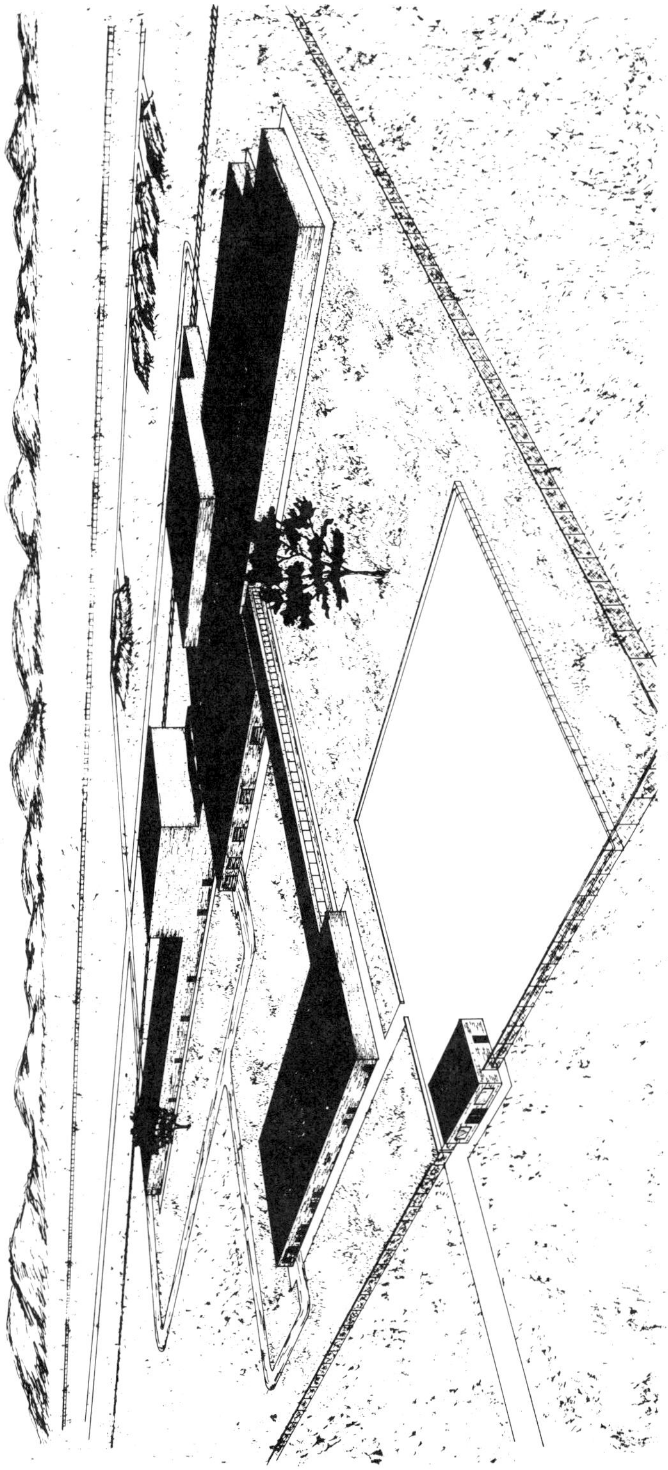

DARCOM Ammunition Center

Figure 1. Standard chemical maintenance facility. Artist's concept.

FACTORS INFLUENCING DESIGN:

Major factors influencing design (Figure 2) were identified as: regulatory criteria, operational capabilities, current state of the art, and agent criteria.

REGULATORY CRITERIA:

Regulatory criteria as provided in the DARCOM Safety Manual, DARCOM Safety Regulations for Chemical Agents, DOD Ammunition and Safety Standards and the Army Chemical Surety Program were found to have a significant impact on facility design.

It should be noted that regulatory and environmental guidance was found primarily pertinent to storage activities or demilitarization operations with limited guidance relevant to maintenance operations.

The Safety Manual AMCR 385-100 prescribes general safety rules for the US Army Materiel Command including those relevant to facility construction for explosive materiel operations and storage requirements, personal protective clothing and equipment together with quantity distance standards of explosives.

Safety Regulations for Chemical Agents as covered in DARCOM-R 385-102 and AMCR 385-31 establishes minimum safety criteria for use in processing, handling, storage, transportation, disposal and decontamination of agents GB, VX, H, HD, and HT.

These regulations also provide more specific guidance relevant to disassembly and remote operations, together with many specifics for personnel protection as air monitoring, protective clothing, showers, and communications. The requirements are also identified for ventilation and filtering of exhaust air to insure no discharge or escape of agent to the atmosphere.

The DOD Ammunition and Explosives Safety Standards DOD 5154.4S describes standards applying to combined toxic chemical and explosive hazards to a maximum credible event and its effect on hazard zone distance calculations. Also within the standards are found safety criteria relating to selection of the type of operational containment required, total versus vapor containment.

The Chemical Surety Program Regulation AR 50-6 prescribes procedures for the safe, secure and reliable life cycle management of chemical agents and munitions. The chemical surety program includes both physical and security design requirements for chemical security structures.

OPERATIONAL CAPABILITIES:

Facility operational capabilities and requirements were established with review of the chemical munitions inventory identifying both current and future maintenance requirements.

In identifying operational criteria, process flows were developed for each item to be processed. Typically, process flows fol-

DARCOM Ammunition Center

Figure 2. Standard chemical maintenance facility. Factors affecting design.

lowed much the same pattern for the majority of the items.

Process flow charts further identified both specific and unique operational and equipment requirements of each item for receiving, unpacking, removal of explosive components, cleaning, and recoating of interior surfaces, abrasive cleaning, and repainting of exterior surfaces. This was followed by the reassembly of explosive components, repacking, and shipping.

Operational layouts were developed providing physical capabilities to accommodate those requirements as defined by the process flow charts. Operational layouts identified both space requirements and requirements for Ammunition Peculiar Equipment (APE), to meet individual item configurations.

In developing conceptual designs for process equipment, it should be noted every effort was taken to protect workers from agent exposure.

Operational philosophy dictated that all operations involving the opening or disassembling of munitions, which could release or expose agent vapor, would be performed remotely within a vapor containment chamber. Prior to removal of items from vapor containment chambers, they would be monitored and verified free of agent.

Following the operational philosophy as described, would provide a safe working environment requiring only Level D or E protective clothing (explosive handlers coveralls with a slung mask) (Figure 3).

The facility requirements for both service and support elements, as utility and environmental systems together with the personnel support area (change house), chemical laboratory and surveillance area were identified and included within the frame work of the design.

CURRENT STATE OF THE ART:

The current state of the art in terms of equipment and support measures had a significant effect on facility design. Of major importance was the availability of agent monitoring and detection devices, protective unitorms for personnel, and acceptable exhaust air filters, together with the capability to handle different types of decon solutions.

Agent exposure limits as established by the Army Surgeon General (Figure 4) for unprotected personnel, dictated the requirement for monitoring and detection systems with a fast response time.

The use of the real time monitor (RTM) with low level sensitivity capability (0.0001 mg/m^3 for GB) and response time of ten minutes together with the M-8 Alarm with a higher level sensitivity (0.2 mg/m^3 for GB) and response time of one minute are planned for use within the facility.

The requirement for use of Level A clothing is anticipated to be at very infrequent intervals, primarily, for handling of leak-

DARCOM Ammunition Center

Figure 3. Personnel protective clothing

AGENT	CONCENTRATION	DURATION
GB	0.001 mg/m^3	1 hour
GB	0.0003 mg/m^3	8 hours
GB	0.0001 mg/m^3	8 hours/day, indefinitely
VX	0.00005 mg/m^3	1 hour
VX	0.00002 mg/m^3	8 hours
VX	0.00001 mg/m^3	8 hours/day, indefinitely
Mustard	0.01 mg/m^3	3 hours
Mustard	0.005 mg/m^3	8 hours
Mustard	0.003 mg/m^3	8 hours/day, indefinitely

DARCOM Ammunition Center

Figure 4. Unprotected personnel exposure limits. Production worker exposure levels.

ers discovered on the line and for area decontamination. To provide personnel with both increased agent protection, mobility, and comfort, the use of the new disposable protective ensemble (DPE Suit) (Figures 5, 6, 7, 8, and 9) developed for chemical demil is planned for use.

At present, air filters capable of meeting current pollution abatement requirements are limited to charcoal or packed column scrubbers. These filters are bulky and expensive, and require continual maintenance to meet operational requirements. In addition, problems exist with the disposal of spent charcoal and by-products from the packed column scrubbers.

AGENT CRITERIA:

Maintenance operations involved with items containing an agent, present the potential use of decontamination solutions. These solutions are highly corrosive and present handling, mixing, and storage problems. There is no one solution that can be utilized for the decontamination of all agents. The decon solutions are not compatible with each other and require methods of assuring the proper decon solution is available for the commodity being worked.

The chemical agents also dictate the requirement for the use of non-porous construction materials, together with epoxy coatings for the floor and walls of operational bays.

Operational equipment with a potential for agent exposure, requires a design capable of decontamination and repair.

FACILITY DESIGN:

The facility design concept as developed, provides an operational capability which satisfies both processing and regulatory requirements as identified.

An artist's concept of the completed facility as shown (Figure 10) contains four major areas.

The personnel support area, primarily a change house and the personnel control center, is connected to the main building by a covered walkway.

The service and support area, centrally located in the main building, provides for operational control and support functions such as: surveillance, decontamination facilities, utility and environmental systems and chemical laboratories.

The explosive wing consists of individual operational bays providing separation of dissimilar operations with explosive dividing walls, for the renovation and maintenance operations requiring disassembly or assembly of explosive components.

The non-explosive wing supports the maintenance and renovation of items not having explosive components.

The facility's general arrangement is as shown (Figure 11). The layout shows the general configuration of the main building

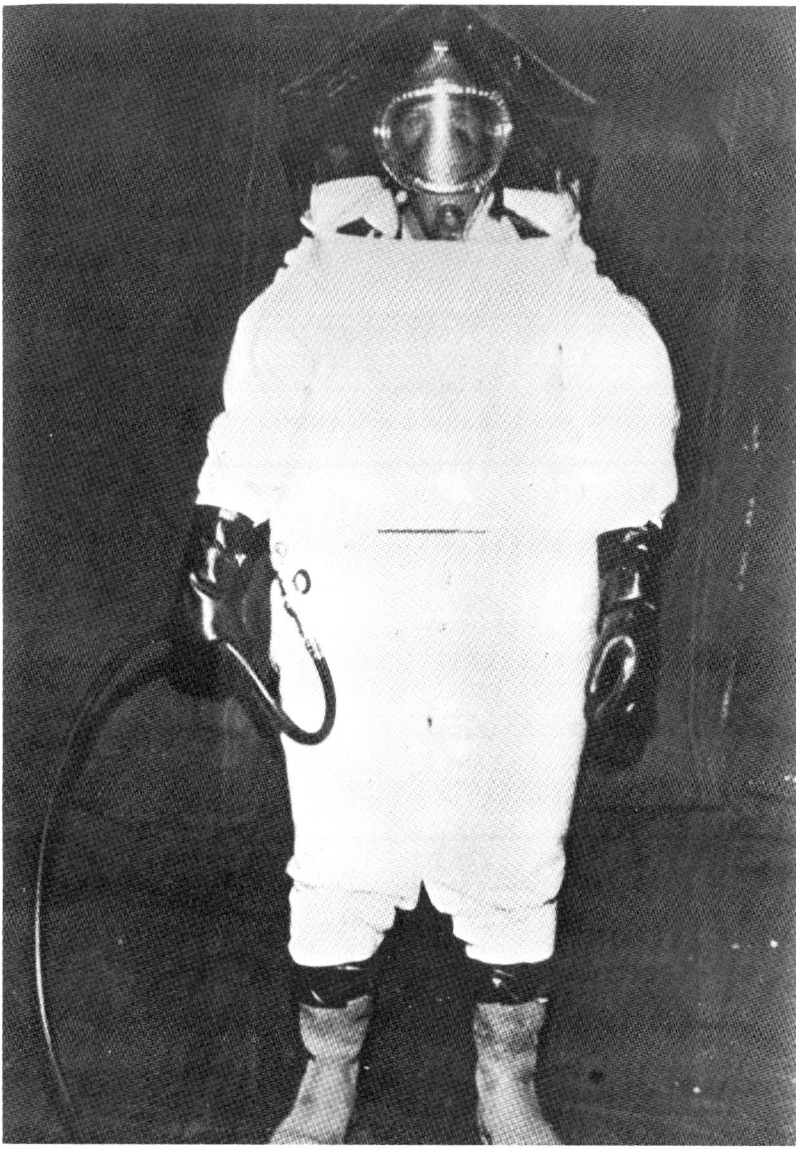

DARCOM Ammunition Center

Figure 5. Disposable protective ensemble (DPE suit)

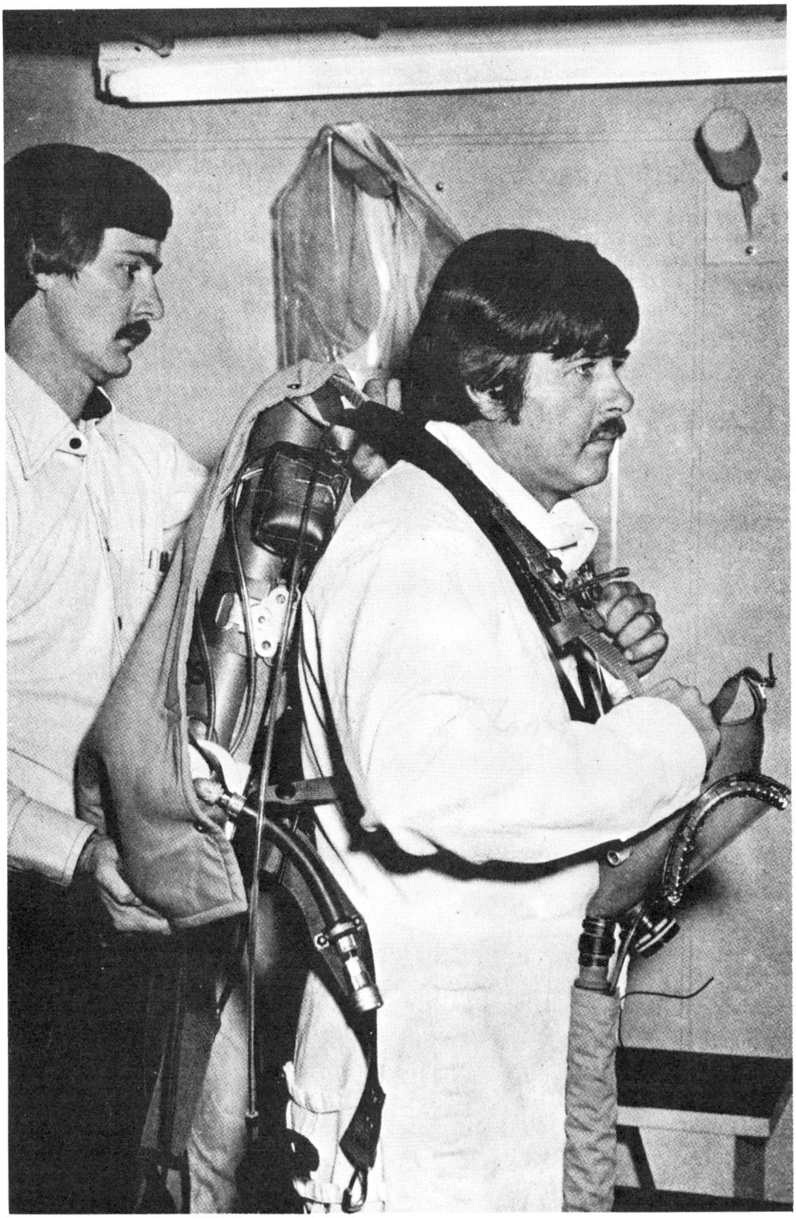

DARCOM Ammunition Center

Figure 6. Disposable protective ensemble (DPE suit)

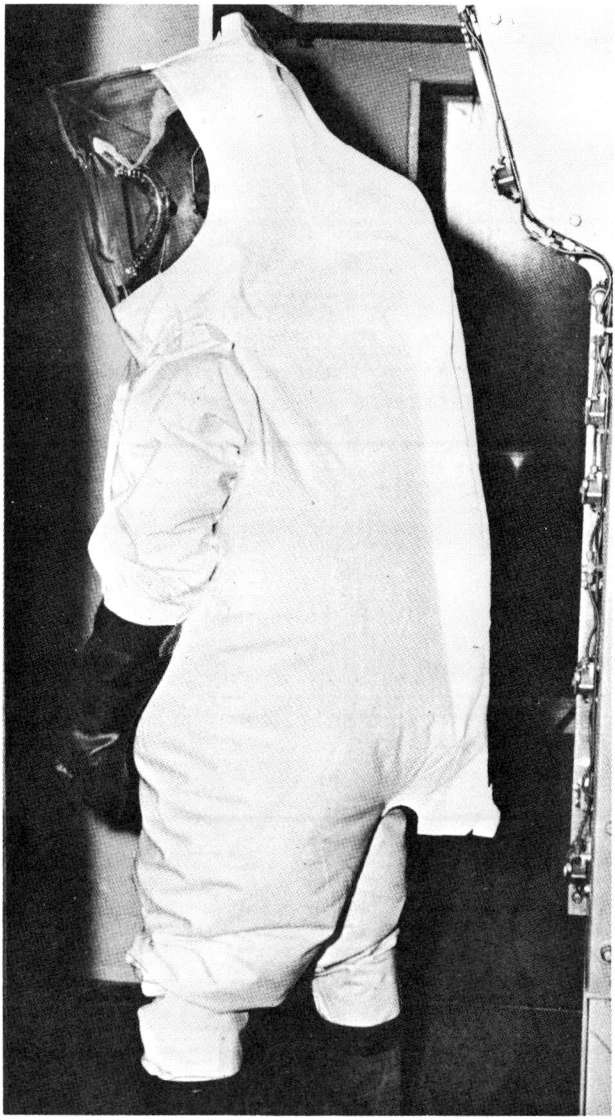

DARCOM Ammunition Center

Figure 7. Disposable protective ensemble (DPE suit)

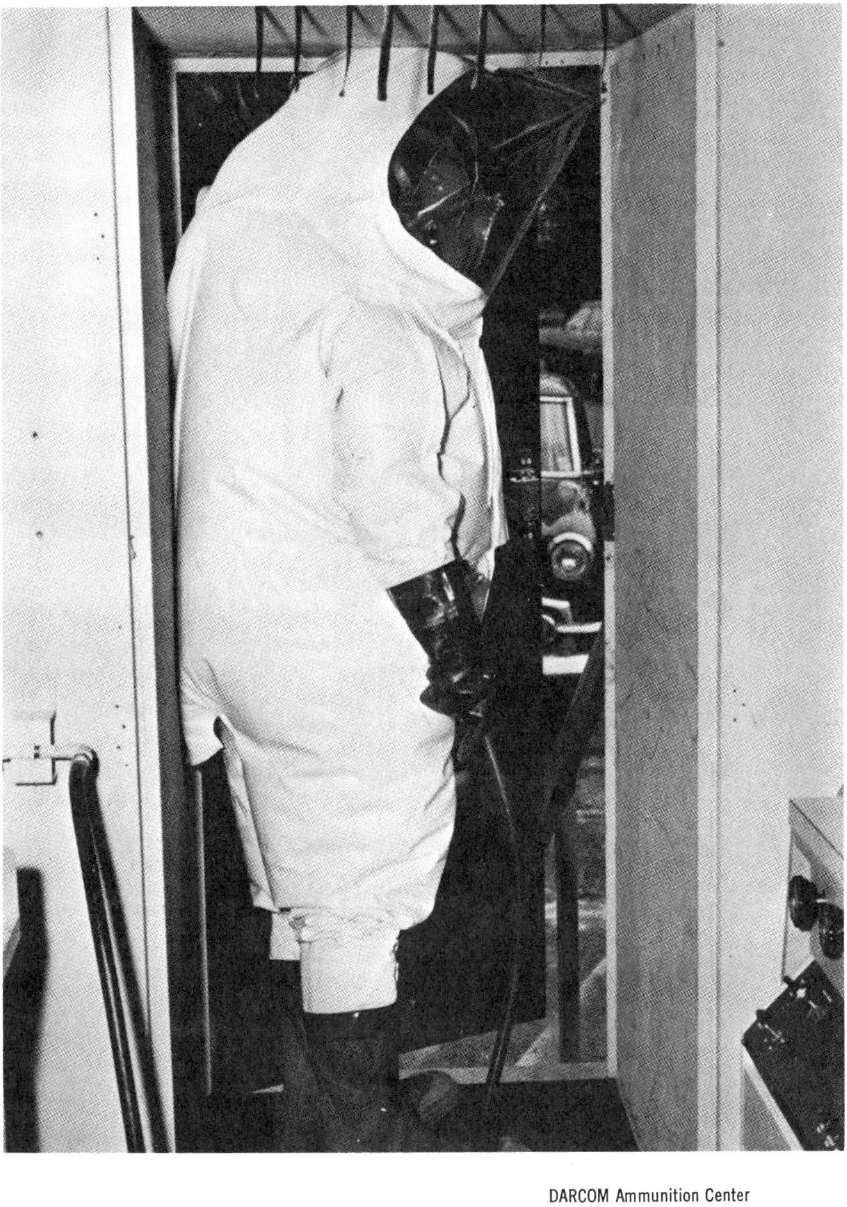

DARCOM Ammunition Center

Figure 8. Disposable protective ensemble (DPE suit)

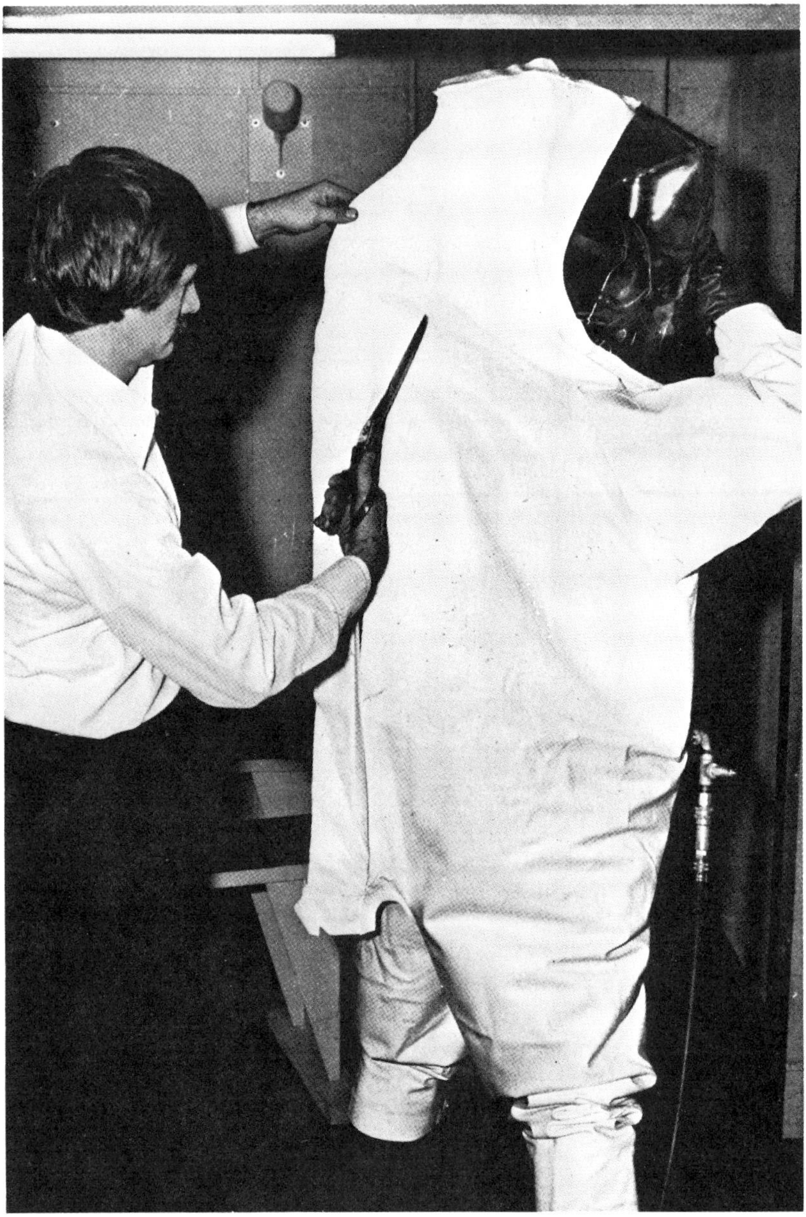

DARCOM Ammunition Center

Figure 9. Disposable protective ensemble (DPE suit)

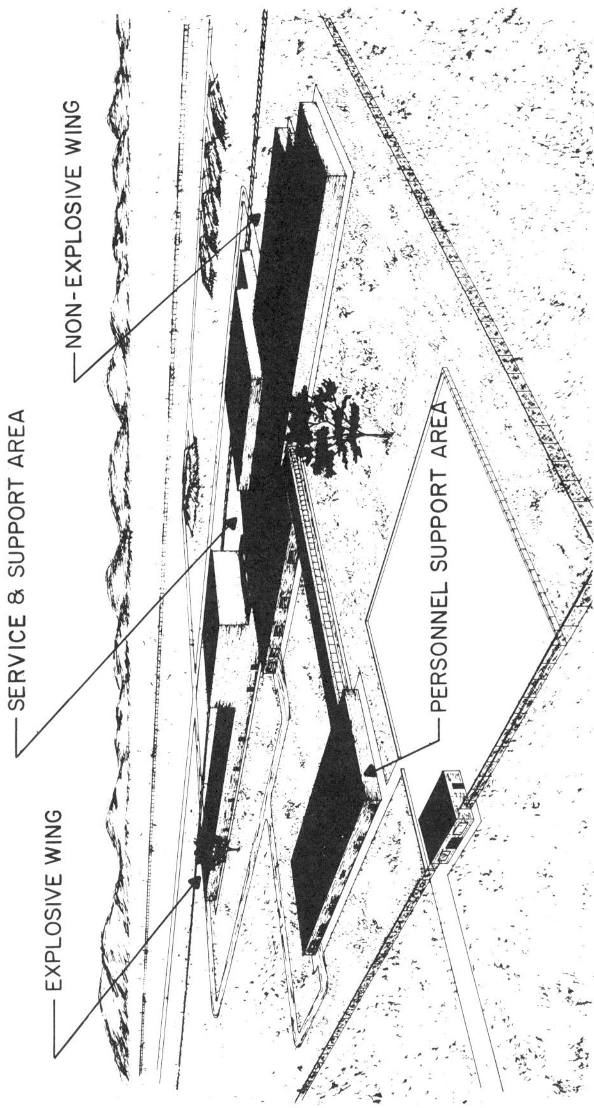

DARCOM Ammunition Center

Figure 10. Standard chemical maintenance facility. Artist's concept.

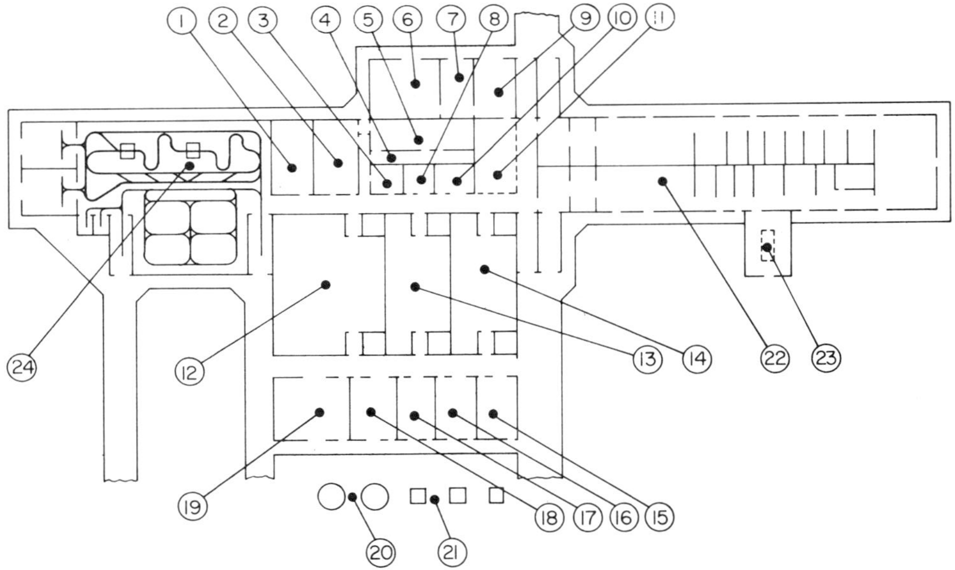

1 Protective Clothing - Doffing Area
2 Protective Clothing - Donning Area
3 Office
4 Control Center Observation Area
5 Control Center
6 Air Conditioning Equipment Room
7 Emergency Generator Room
8 Facility & Equipment Support Area
9 Air Compressor Room
10 Facility Equipment Storage
11 Paint Area - Dunnage
12 Chemical Laboratory
13 Decontamination Area
14 Surveillance Area
15 Decon Storage & Mixing Room
16 Decon Sump Room
17 Decon Dryer Room
18 Incinerator
19 Boiler Room
20 Scrubber Systems
21 Electrostatic Precipitators
22 Explosive Operational Area
23 Explosive Containment Chamber (ECC)
24 Non-Explosive Operational Area

DARCOM Ammunition Center

Figure 11. Standard facility for maintenance, preservation and packaging, surveillance and assembly of chemical munitions

with the explosive wing on the right, the non-explosive wing on the left, and the support area in the center.

PERSONNEL SUPPORT AREA:

The personnel support area layout is as shown (Figure 12). The layout of the change house provides for an orderly flow and control of personnel having a security control point, locker room with shower facilities, lunch room, clothing/equipment issue, and storage area. In addition, medical facilities required to support a chemical operation include an ambulance.

SERVICE & SUPPORT AREA:

The service and support area as shown (Figure 13) is located in the center of the main facility providing service to both operational wings.

Within the service and support area are found a personnel service area which includes: facilities for donning and doffing of DPE suits, office and operational control center, air conditioning system, emergency generator and air compressors for plant and life support air systems.

The paint area provides necessary equipment for renovation, painting, and restencilling of dunnage and packaging materials, separating this work area from potentially more hazardous areas.

The facility equipment support and storage areas are provided for service and repair of operational and material handling equipment.

The chemical laboratory, decontamination and surveillance areas as shown have a potential for agent exposure and therefore are provided with air lock separation.

The dryer room, incinerator and decontamination facilities provide the capability for detoxification and disposal of spent decon.

The pollution abatement systems, including scrubbers and electro-static percipitators, etc., assure that no agent or hazardous materials are released to the atmosphere.

EXPLOSIVE WING:

The explosive wing is as shown on the right side of the general arrangement (Figure 11).

The explosive wing utilizes individual bays for maintenance operations that can isolate an area contaminated by a leaker or agent spill. Although no explosions are expected to occur in this facility, explosive blast walls are incorporated to permit facility utilization during its 25 year life expectancy. The design of the explosive blast walls is in accordance with TM 5-1300 for 1,000 pounds of net explosive weight (NEW) per bay requiring a wall thickness of 30 inches.

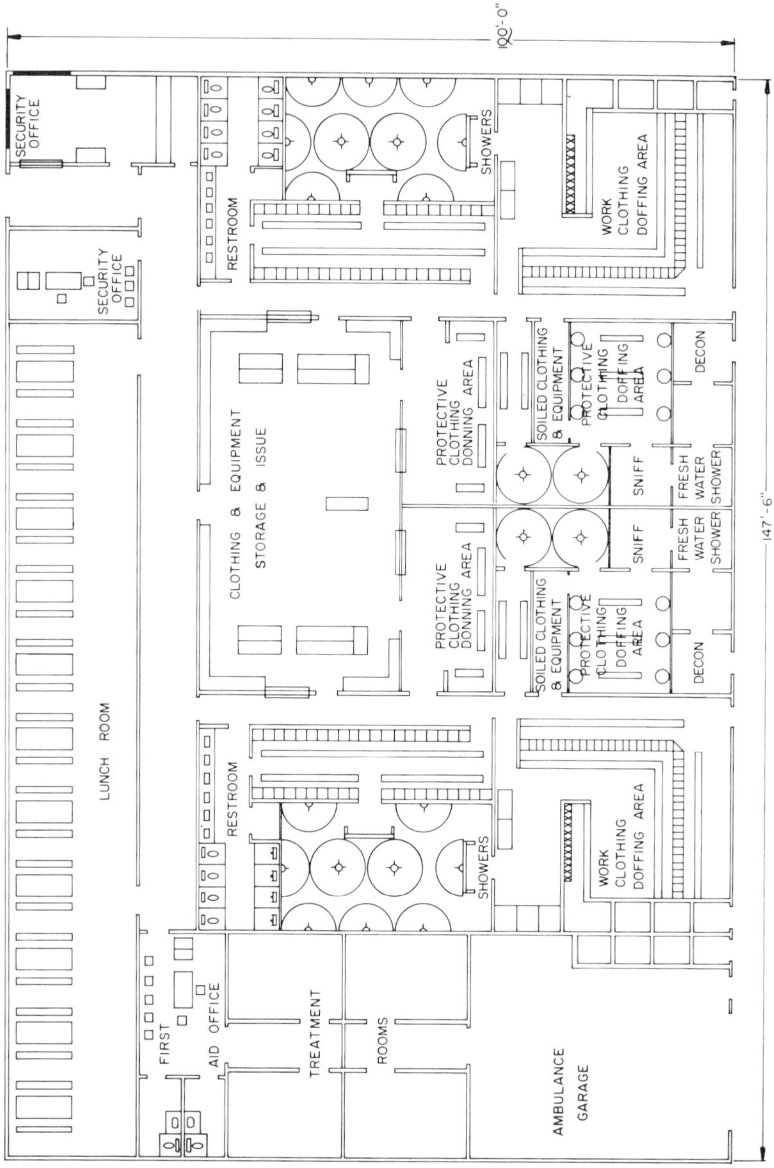

DARCOM Ammunition Center

Figure 12. Standard chemical maintenance facility. Change house.

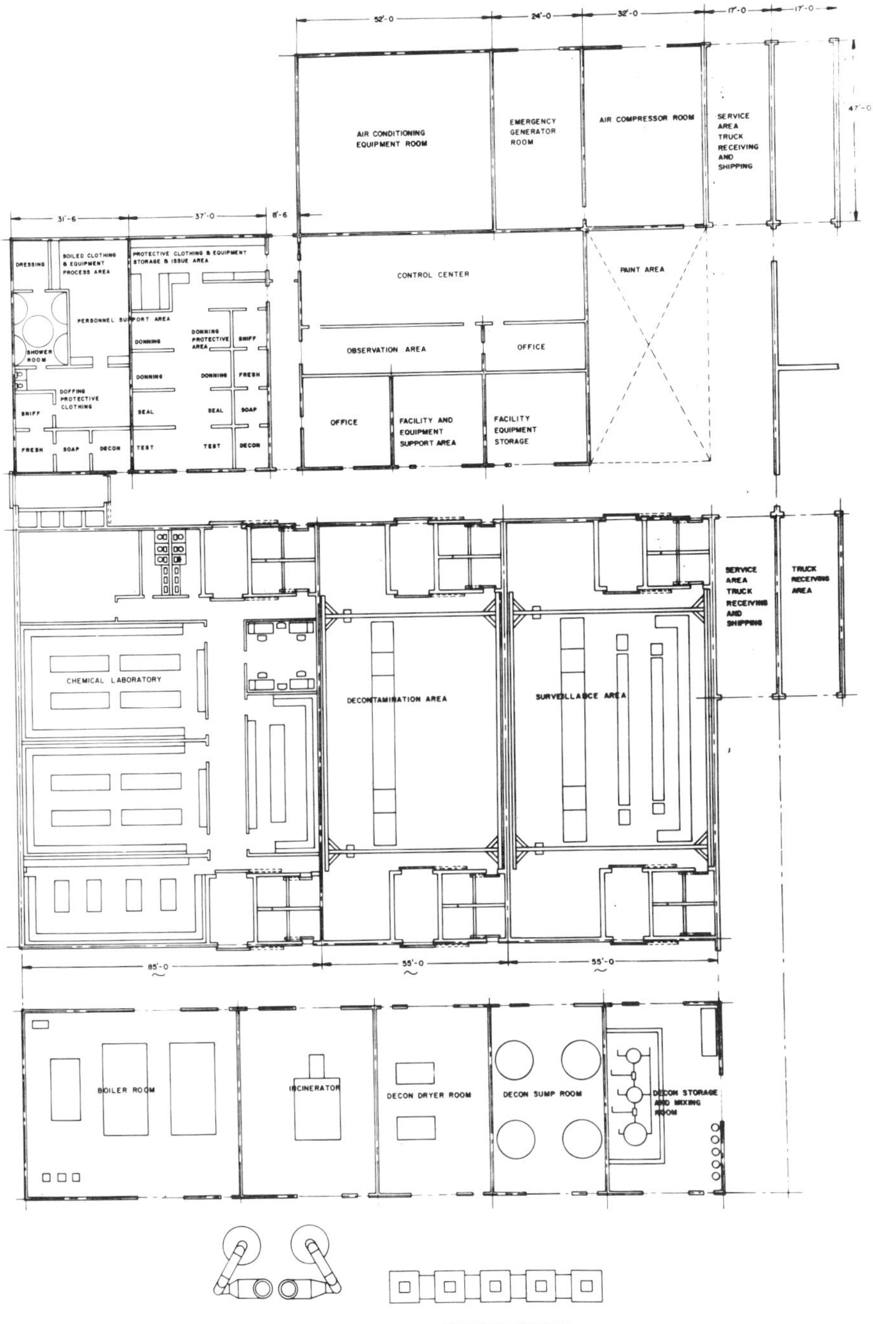

DARCOM Ammunition Center

Figure 13. Standard chemical maintenance facility. Service and support area.

An explosive containment chamber (ECC), is as shown attached to the explosive wing that will permit continuity testing of rockets, as well as the disassembly of explosive components from agent filled munitions requiring undue force. The operational sequence cycle of the ECC provides for total interlock control assuring complete vapor and explosive containment during performance of remote operations. Prior to the start of any operation or test the interlock control requires closing, sealing and locking of the door including closing and sealing off of the ventilation system. Following completion of an operation cycle interlock control requires opening of the ventilation system, purging and monitiring the atmosphere prior to opening of the door and removal of materials.

Vapor containment only is provided for all operations performed within the explosive wing assuring no escape of agent vapor to the outside atmosphere.

The general layout of the explosive wing is as shown (Figure 14). The large bays on the left are for unpack and repack operations with individual maintenance operations being performed in the smaller bays.

A typical layout of the unpack bay is as shown (Figure 15). The area to the left provides for the control and containment of agent through the use of air locks, while the remaining area of the bay is devoted to unpacking equipment.

Personnel working in this area may work in Level D or E protective clothing.

Typical maintenance operations performed under controlled environments are as shown (Figures 16, 17, 18, 19, and 20).

NON-EXPLOSIVE WING:

The non-explosive wing is as shown on the left side of the general arrangement (Figure 11). This wing is designed primarily to handle large bulky items having no explosives assembled.

The layout of the non-explosive wing is as shown (Figure 21). The power and free conveyor system provides process operational flexibility for maintenance operations associated with abrasive cleaning, inspection, painting and shipping of large items.

The design and layout of the chemical maintenance facility provides for both operational flexibility and a capability to support process requirements as identified together with a safe working environment. The protection of workers from agent exposure is provided with an air lock separation between operational areas, preventing spread of agent vapor together with agent detection equipment positioned throughout the facility to monitor the environment.

FACILITY COST ESTIMATE:

The facility (Figure 1) was sited adjacent to the chemical

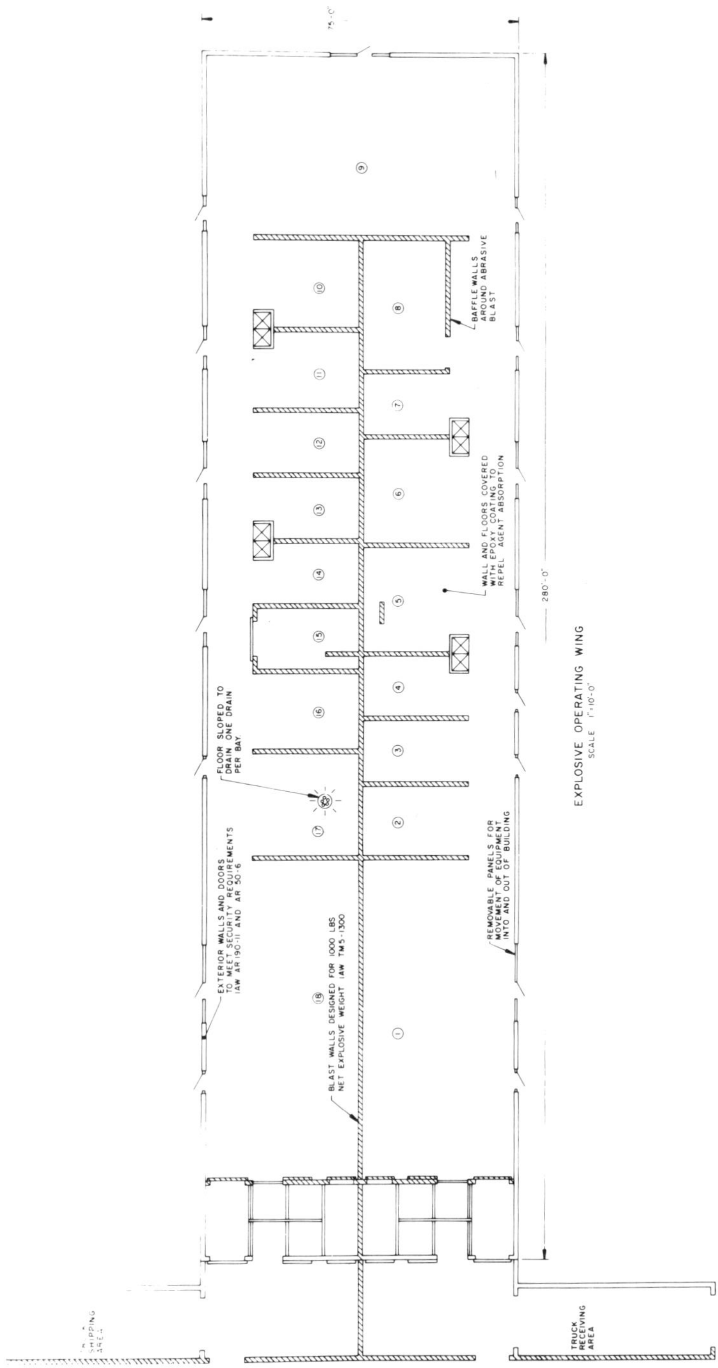

DARCOM Ammunition Center

Figure 14. Standard chemical maintenance facility. Explosive wing.

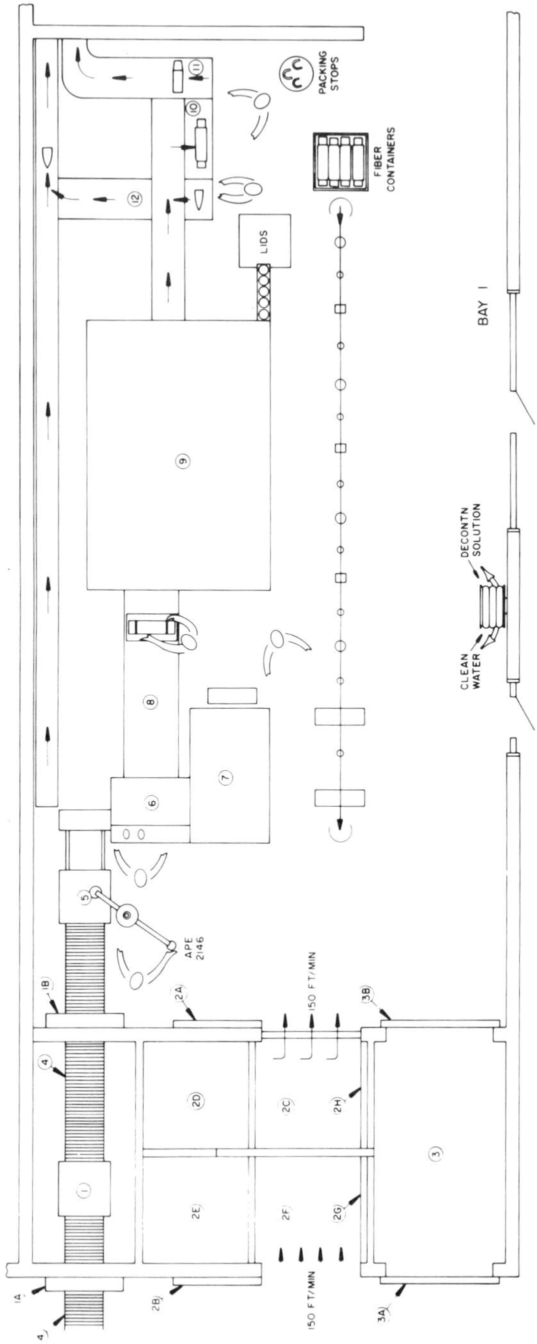

DARCOM Ammunition Center

Figure 15. Standard chemical maintenance facility. Explosive unpack bay.

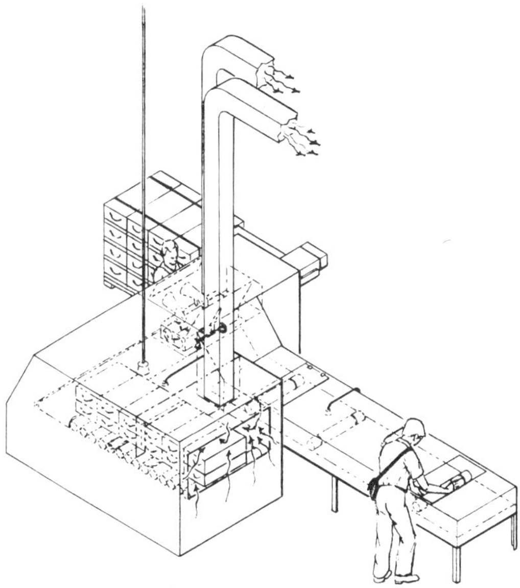

DARCOM Ammunition Center

Figure 16. Standard chemical maintenance facility. Fiber container unpack concept.

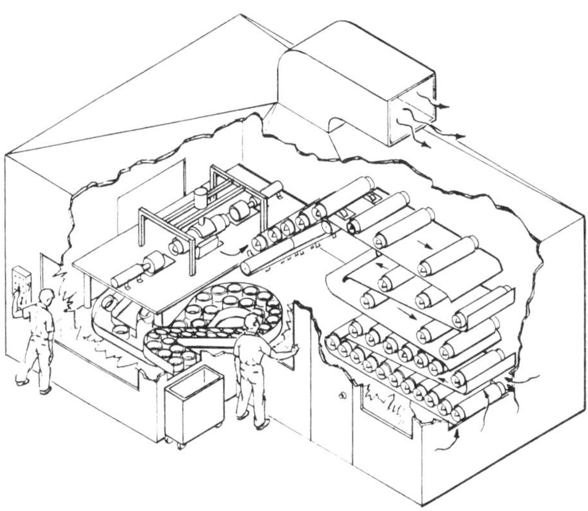

DARCOM Ammunition Center

Figure 17. Standard chemical maintenance facility. Lid removed in vapor containment chamber.

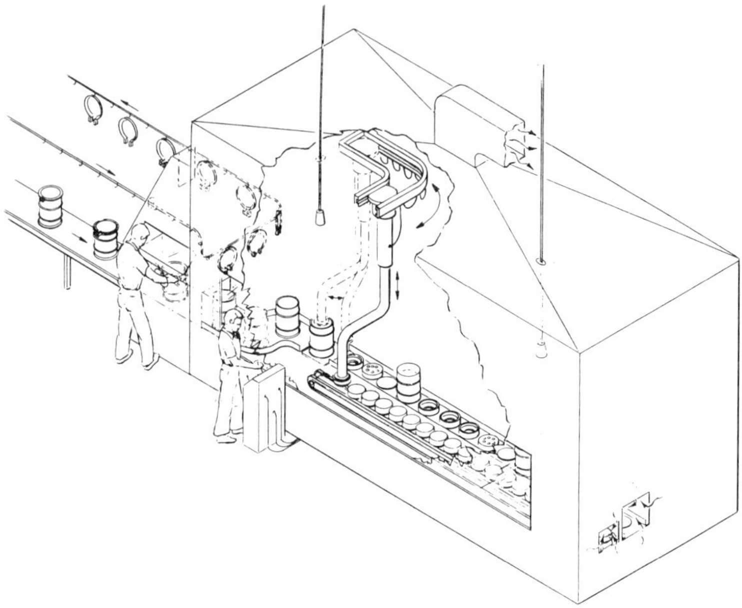

DARCOM Ammunition Center

Figure 18. Standard chemical maintenance facility. M-23 mine unpack station.

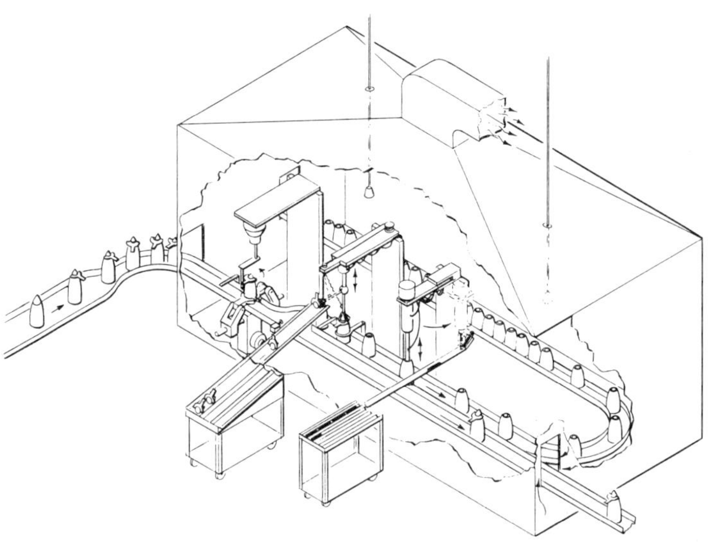

DARCOM Ammunition Center

Figure 19. Standard chemical maintenance facility. Fuze and Burster removal station.

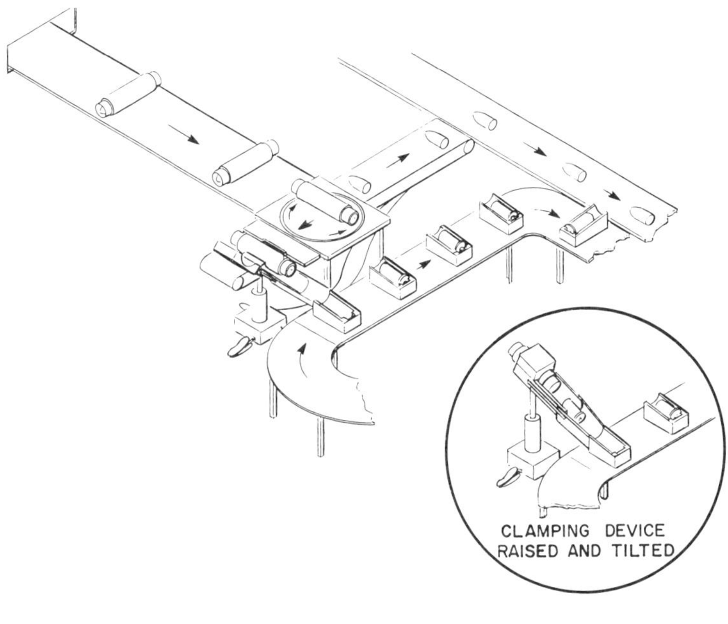

DARCOM Ammunition Center

Figure 20. Standard chemical maintenance facility. Fiber container unpack station.

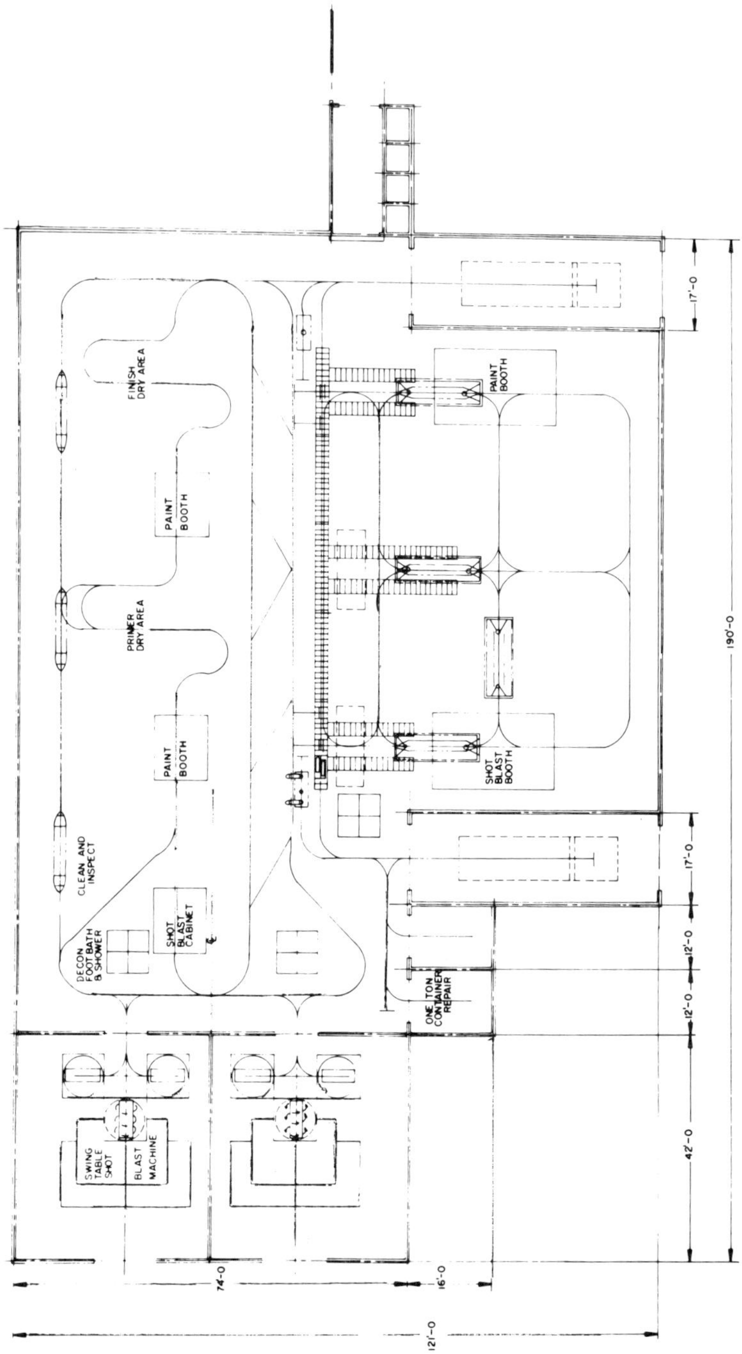

DARCOM Ammunition Center

Figure 21. Standard chemical maintenance facility. Nonexplosive wing.

DESCRIPTION	QUANTITY	UNIT	UNIT PRICE	TOTALS ($000)
1. MAIN. BUILDING				
A. EXPLOSIVE AREA				
GENERAL CONST.	24,200	SF	78.00	1,887.6
MECHANICAL	24,200	SF	228.48	5,529.7
ELECTRICAL	24,200	SF	40.35	976.5
SUBTOTAL	24,200	SF	346.83	8,393.8
B. NON-EXP AREA				
GENERAL CONST.	22,668	SF	40.00	906.7
MECHANICAL	22,668	SF	228.17	5,172.2
ELECTRICAL	22,668	SF	16.90	383.1
SUBTOTAL	22,668	SF	285.07	6,462.0
C. SERVICE & SUPPORT AREA				
GENERAL CONST.	56,650	SF	44.65	2,529.4
MECHANICAL	56,650	SF	45.60	2,583.0
ELECTRICAL	56,650	SF	32.80	1,858.1
SUBTOTAL	56,650	SF		6,970.7
TOTAL MAIN. BLDG.	102,178	SF		21,826.5
CHANGE HOUSE	14,971	SF	86.00	1,287.5
MAGAZINES (40 X 25)	4	EA	152,000	608.0
MAGAZINE (2240 SF)	1	EA	286,720	286.7
ECC STRUCTURE	1	EA	L. S.	250.0
TOTAL				2,432.2
GRAND TOTAL				$24,258.7

DARCOM Ammunition Center

Figure 22. Standard facility for maintenance, preservation and packaging, surveillance and assembly of chemical munitions. Cost estimate.

DESCRIPTION	TOTALS ($000)	DESCRIPTION	TOTALS ($000)
1. STRUCTURES	24,258.7	d. DRAINAGE PIPE (48")	24.0
2. CLEARING & GRUBBING	84.0	e. HEADWALLS OR INLETS	76.8
3. EARTHWORK	398.4	7. GRASSING	57.6
4. FLEXIBLE PAVEMENT		8. WATERLINE	349.4
a. SURFACE COURSE (2")	134.4	9. SEPTIC TANK	16.0
b. BASE COURSE (6")	94.5	10. RAILROAD	224.0
c. SUBBASE (6")	100.8	11. 150,000 GAL WATER TANK	348.0
5. RIGID PAVEMENT		12. FENCING	99.0
a. P. C. CONCRETE (7")	193.2	13. MECHANICAL EXT	257.1
b. SUBBASE (6")	35.9	14. ELECTRICAL EXT	876.1
6. STORM DRAINAGE		15. HELI PAD	28.8
a. DRAINAGE PIPE (18")	9.4	TOTAL SITING COST	3,441.1
b. DRAINAGE PIPE (30")	2.9	TOTAL FACILITY COST	$27,697.8
c. DRAINAGE PIPE (42")	28.8		

DARCOM Ammunition Center

Figure 23. Standard facility for maintenance, preservation and packaging, surveillance and assembly of chemical munitions. Site preparation cost estimate.

	$ THOUSAND		
	FACILITY	EQUIPMENT	TOTAL
ANNISTON ARMY DEPOT	21235.8 29629.6	7996.6 10508.8	29232.4 40138.4
LEX-BLUE GRASS	22638.6	4868.1	27506.7
PINE BLUFF ARSENAL	26743.8	5971.1	32714.9
PUEBLO DEPOT ACTIVITY	25989.8	4778.1	30767.9
TOOELE ARMY DEPOT	29322.0	7292.3	36614.3
UMATILLA DEPOT ACTIVITY	30692.1	6243.1	36935.2
TOTAL	156622.1 165015.9	37149.3 39661.5	193771.4 204677.4

Figure 24. Standard facility for maintenance, preservation and packaging, surveillance and assembly of chemical munitions including support and process equipment. Total cost estimate.

storage area at installations involved. Specific siting requirements were identified which considered: the availability and extension of utility services, roads and railroad sitings. Siting information was provided to the Huntsville District Corps of Engineers which served as a basis for developing site preparation costs.

Based on the data provided by the installations involved and the design data developed for the facility, the Huntsville District Corps of Engineers assembled a team of estimators at the DARCOM Ammunition Center and prepared facility cost estimates.

The basis for construction cost estimates were for the facility to be in the FY80 MCA program, with construction costs escalated to reflect January 1971 time frame, as the midpoint of construction. All estimates for the facility were based on unit pricing at Anniston and factored for other sites.

The cost estimates were developed as shown (Figure 22): the explosive wing $8,400,000; a non-explosive wing $6,500,000; and the service and support area approximately $7,000,000 for a total of $21,900,000 for the main building. The change house, four service magazines and a large conditioning building add another $2,400,000 for a total of $24,300,000.

The site preparation cost, as developed for data provided, for Anniston is $3,400,000 (Figure 23).

The total estimated cost for the proposed facility sited at Anniston Army Depot including equipment is $27,700,000.

With the development of a modular designed facility, it was possible to tailor the facility to meet the individual requirements of each installation.

This chart (Figure 24) reflects the combination of the facility cost estimate together with the equipment cost estimate tailored to stocks currently located at the installations. The deletion of the non-explosive wings from the Anniston, Lexington-Blue Grass and Pueblo facilities is reflected in the facility and equipment costs.

The total estimated cost for tailored standard chemical maintenance facilities constructed at each installation is $193,700,000 with one explosive wing at Anniston or $204,700,000 for two explosive wings at Anniston.

CONCLUSIONS:

The development of the standard maintenance facility design as presented (Figure 1) concludes that both operational and environmental requirements can be accommodated. It satisfies a real need for a facility that will provide a safe and productive environment for the maintenance of toxic chemical munitions. Also that safety, security, and environmental criteria are stringent but the facility design as presented will comply with those known requirements.

Literature Cited:

1. Safety Manual, AMCR 385-100, HQ US Army Materiel Command, Alexandria, VA, DARCOM Change 3, 11 January 1977.

2. Safety Regulations for Chemical Agents GB and VX, DARCOM-R 385-102, HQ US Army Materiel Development and Readiness Command, Alexandria, VA, 28 February 1977.

3. Safety Regulations for Chemical Agents H, HD, and HT, AMCR 385-31, HQ US Army Materiel Command, Alexandria, VA, 26 September 1975.

4. DOD Ammunition and Explosives Safety Standards, 5154.4S, Interim Change 2-1, 15 July 1976.

5. Chemical Surety Program, AR 50-6, HQ Department of the Army, Washington, DC, 5 November 1976.

6. Structures to Resist the Effects of Accidental Explosions, TM 5-1300, Department of the Army, Washington, DC, 15 June 1969.

RECEIVED November 22, 1978.

18

Development of Highly Sensitive Monitors for the Detection of Anticholinesterase Compounds

DONALD C. BEHRINGER, FREDERICK C. BALADUF, and VINCENT N. CAMMARATA

Chemical Systems Laboratory, Aberdeen Proving Ground, MD 21010

Two highly sensitive monitors for the detection of anticholinesterase compounds were developed. One monitor was specifically developed to meet stringent time concentration requirements established by the Surgeon General of the Public Health Service and provides rapid and continuous detection of trace quantities of compounds of the type noted in Figure 1. As will be seen, however, the monitors could be useful devices for other enzyme-inhibiting compounds. The monitors will be evaluated by the Army's Project Manager for Chemical Demilitarization and Installation Restoration at the new Chemical Agent Munitions Disposal System (CAMDS), Tooele Army Depot, Utah, during execution of the program to destroy obsolete stocks of toxic compounds. The rapid monitor was developed to automatically monitor exhaust stacks and work areas while the other monitor is a personal dosimeter to be worn by plant workers.

The well-known enzyme colorimetric method of analysis is employed as the basis for the determination of the anticholinesterase compound. The enzyme selected for determination of the presence of the compound is an acetylcholinesterase. As shown in Figure 2, without an inhibitor present, the series of reactions are: The acetylthiocholine (ATCI) is hydrolyzed by the enzyme acetylcholinesterase to produce acetic acid and thiocholine. The thiocholine then reacts with 5,5′-dithio-bis-2,2′-nitrobenzoic acid (DTNB) to form the colored anion, 5-thio-2-nitrobenzoate, production of which is monitored colorimetrically. With an inhibitor present, the enzyme hydrolysis of the ATCI does not occur resulting in a reduction or loss of color which is the basis of the sensitivity of the monitor. As such, this device is a measure of the toxicity of the enzyme inhibitors. A small amount of a highly toxic compound or a larger amount of a less toxic compound will produce similar results. As shown in Figure 3, the analytical reactions are performed in two glass coils immersed in a water bath at 32°C. In the first coil, the sample, enzyme, and segmentation air are introduced. If agent is present in the sample, inhibition will occur. The contact time is about 3 minutes. The stream then flows to the second coil where the substrate, ATCI, and the color developer, DTNB, are introduced. The hydrolysis of the substrate proceeds to yield acetic acid and thiocholine iodide (TCI). As the TCI is produced, it reacts with the DTNB to form the colored anion, 5-thio-2-nitrobenzoate. The reaction stream is then pumped to the colorimeter. The time for this stage is 2 minutes 40 seconds. The total time from

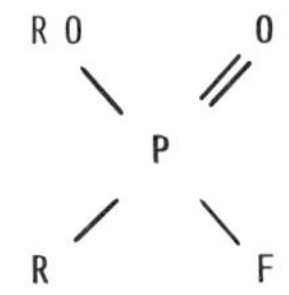

Figure 1. Anticholinesterase contaminant

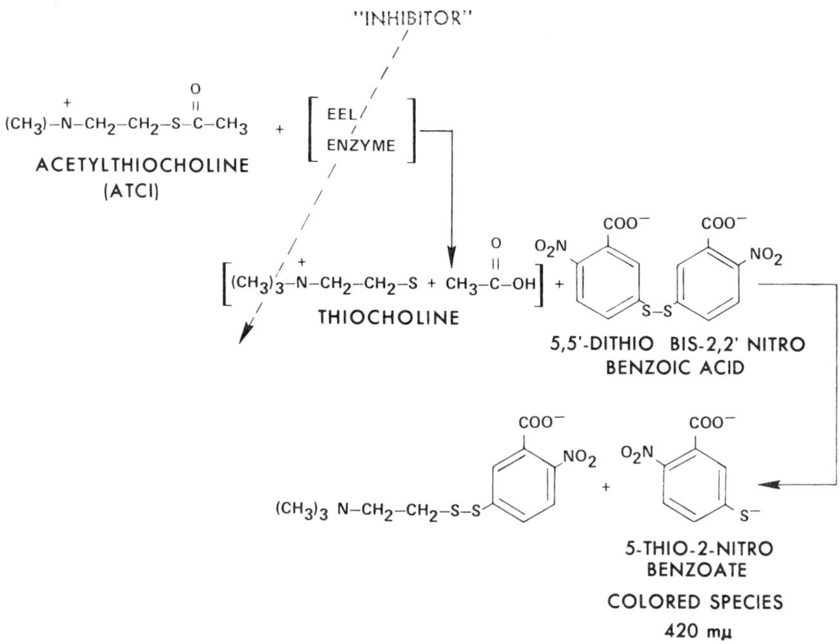

Figure 2. Monitor detection concept

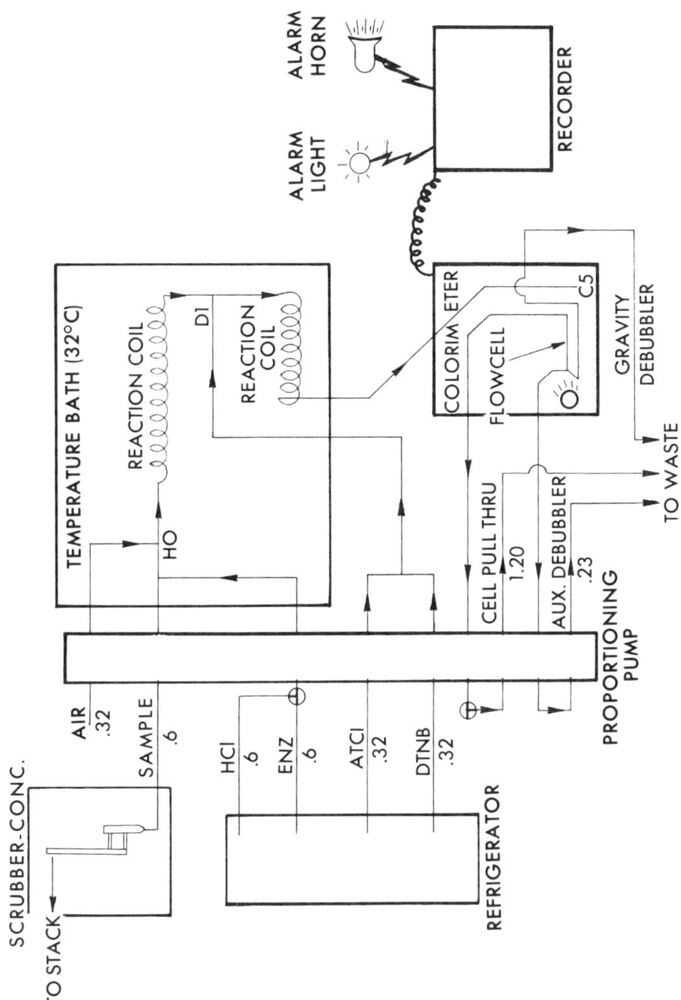

Figure 3. Monitor system

probe inlet to colorimeter is about 7 minutes. The agent response time varies from 8-1/2 to 11-1/2 minutes depending upon at what point in the 3-minute cycle the agent is first absorbed.

The analysis is completed by a colorimeter that electronically compares the color intensity of the analytical stream against an air reference. The ratio signal is sent to a strip chart recorder and to a counter and logic card. Activation of visual and audible alarms will occur if predetermined agent levels exist.

The analytical precision and accuracy are controlled by a peristaltic pump that meters out exact amounts of the various reagent solutions. The flow rates for the absorbent liquid, probe sample, enzyme, acid, ATCI, and DTNB, as well as the waste products, are all precisely controlled so the analytical process can be continually repeated. The color thus formed in the absence of inhibitor and the color loss when inhibitor is present are always predictable (±1%).

A timer-activated solenoid valve replaces the enzyme with dilute acid (HCl) for 27 seconds during each 180-second cycle. The use of the acid wash removes all residual color that was produced during the previous 3-minute cycle. We found that this step was necessary to avoid the troublesome color "buildup" that is encountered in most automated continuous systems where color is formed.

A block diagram of the system is shown in Figure 4 to further indicate the air, solution, and readout processes.

A typical recorder printout is shown in Figure 5. Color is produced by uninhibited enzyme and extinguished by acid during each cycle. The colorimeter converts the color into an electric signal which is received by the record where the signals are manifested as a series of dark and light peaks. The counter and logic electronics evaluate the signals and will activate the alarm systems if the amount of color does not meet a preset level during any cycle. The color of the solution resulting from the reaction is inversely proportional to the concentration of inhibitor in the air sample. Three typical responses resulting from three exposures of the system to an inhibitor are shown in Figure 5. It should be noted that very good baseline stability has been demonstrated during evaluation of the system in the polluted atmospheres found in plant exhaust stacks. We have demonstrated the sensitivity of the system by introducing the low levels of the inhibitor while the system has sampled plant or stack atmospheres. Calibration of the monitor is performed using standard solutions of the compounds of interest. Reagents utilized in the system are shown in Figure 6. Servicing on 7-day intervals is possible using the amounts as indicated. The solutions are kept in a refrigerator which is part of the system. The solutions are stable and usable for periods of at least 2 weeks after preparation if kept cold. With this capability, the monitor can be operated intermittently without the need to continually prepare fresh solutions. A feature of the device is that the refrigerator continues to operate and maintain temperature even when the monitoring functions are not operating.

The system as developed is shown in Figure 7. The device operates from standard 110 v ac; it weighs 450 pounds and measures 6 by 2.5 by 2.5 ft. Shown is the strip chart recorder, refrigerator, and alarm-light indicator.

A unique air-sampling probe shown in Figure 8 was developed to eliminate sampling problems. The absorption and concentration of low volatile compounds had been a particular problem due to the propensity of the materials to adhere to

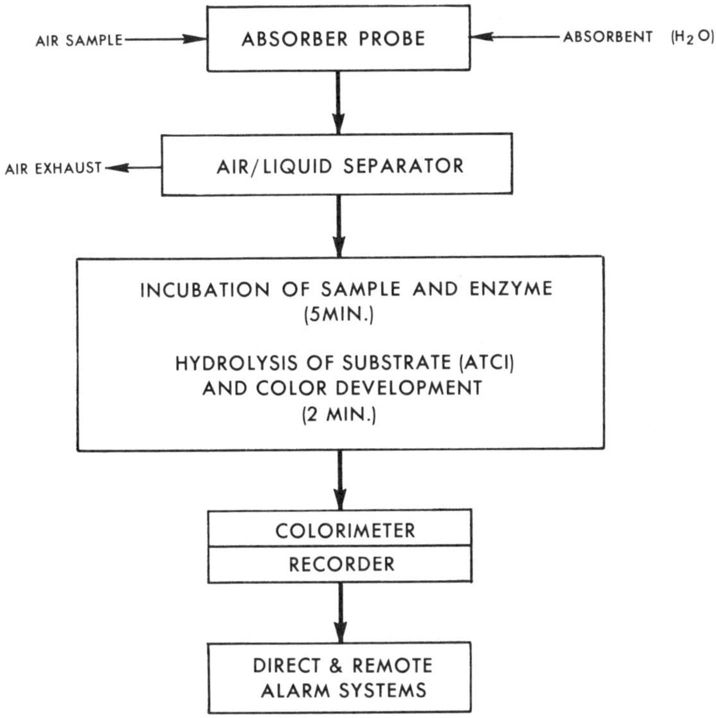

Figure 4. Monitor block system

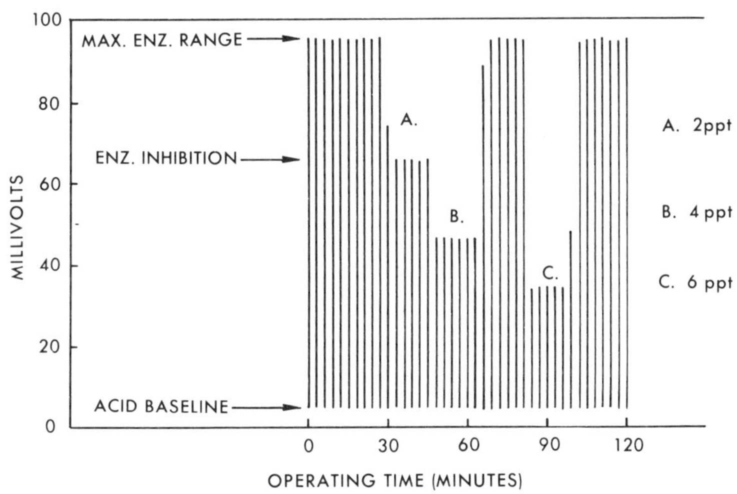

Figure 5. Inhibitor response

REAGENT	FLOW RATE (mls/min)	DAILY REQUIREMENT (liters)	7 DAY WEEK REQUIREMENT (liters)
ATCI .0008M IN DISTILLED WATER	0.32	0.461	3.226
DTNB .0004M IN .05 M TRIS BUFFER (pH 7.4)	0.32	0.461	3.226
HCl 0.2N	0.60	0.130	0.907
ENZYME 50μ/ℓ IN .05 M TRIS BUFFER (pH 8.4)	0.60	0.734	5.141
WATER FOR PROBE FEED	2.00	2.880	20.160

Figure 6. Monitor reagents

Figure 7. Work area monitor

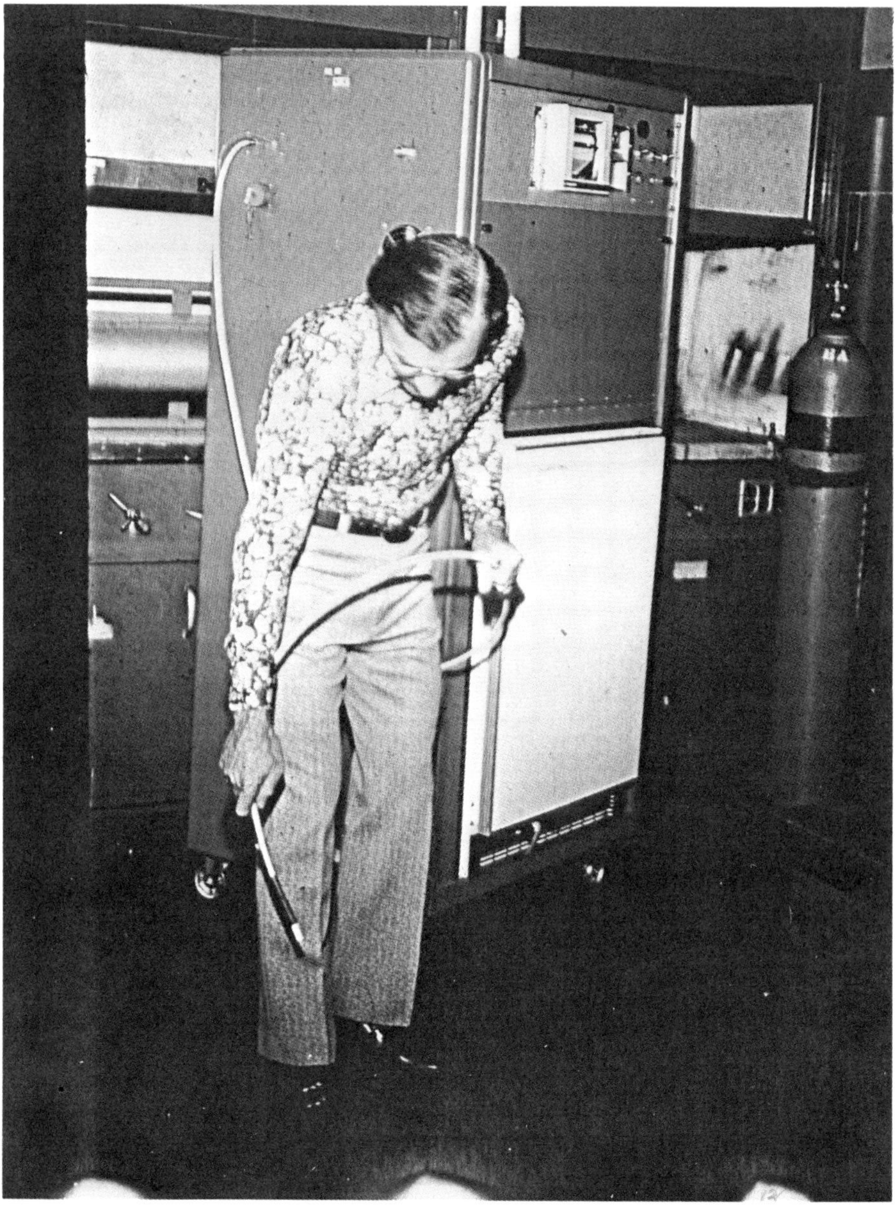

Figure 8. Monitor and sample probe

sampling lines or any internal equipment involved in the transfer of the vapor to the monitor. The air-sampling device used with the subject monitor minimizes the above problems. The device is simply two flexible tubes: one tube draws the air sample and the other tube supplies an absorbent liquid at the air inlet of the first tube. The air and liquid make contact and are drawn concurrently along the inside wall where absorption occurs. The air/liquid stream then empties into a calming chamber which reduces the velocity, allowing the liquid to gravitate into a reservoir where it is available for continuous analysis. In practice, the air/liquid contact is achieved as the liquid "wets" the tube wall surface with a thin-moving liquid layer. The large surface area of the liquid layer presents adequate opportunity for intimate contact and efficient absorption.

The development of the personal dosimeter involved the design, construction, testing, and performance evaluation of a system which fulfills these specific requirements: (1) designed for lightweight and unencumbered attachment to the worker; (2) presents no danger to the health, safety, or welfare of the worker; (3) samples ambient air in the worker's breathing zone in the temperature range of 32° to 105°F for 8 hours; and (4) generates quantitative determinations of individual exposures.

The requirement for sensitivity of the dosimeter system is to detect a concentration of 18 parts per trillion in 8 hours. At this safe limit concentration, the bubbler will collect 1.5×10^{-9} g of contaminant when sampling air at 30 cc/min for an 8-hour period. A sufficient amount of contaminant in solution must be retained for analysis.

The dosimeter design concept follows the principle of commonly used sampling devices which extract trace amounts of toxic substances from air samples drawn through a liquid absorber by solution. The sample collected is sent to the laboratory and the liquid is analyzed by a suitable quantitative procedure such as the colorimetric method just described. The essential components of the dosimeter are an aeration bubbler, a sampling pump, and an interconnecting plastic sampling line. Miniature connectors at each end provide a positive air seal and a "quick-disconnect" feature. A carrier is provided for wearing the dosimeter. The pump is located on the waist belt and the shoulder strap contains a pouch for placing the bubbler in the breathing zone. The dosimeter configuration is shown in Figure 9. Figure 10 shows the dosimeter on the wearer. The bubbler is a miniaturized two-piece glass unit. The lower piece is the aeration chamber containing the absorbent solution which is pH 3.7 distilled water and sulfuric acid, unbuffered. The top piece is an air chamber designed to prevent migration of the solution from the aeration chamber. A "trap" feature built into the air chamber provides confinement of the solution within the bubbler in any position or attitude. The two-piece construction facilitates the loading and unloading of the absorbent and simplifies the handling and analytical procedures in the laboratory.

Extensive field testing determined that the most suitable and accurate pump for the dosimeter application was one designated "Accuhaler 808," which is manufactured by MDA Corporation. The principle of operation is based upon an aspiration cycle which draws a constant volume of sample air through a limiting type of orifice for each stroke of the pump. A counter reads out the pump strokes and the total sample volume is computed for a given sampling period. The dosimeter

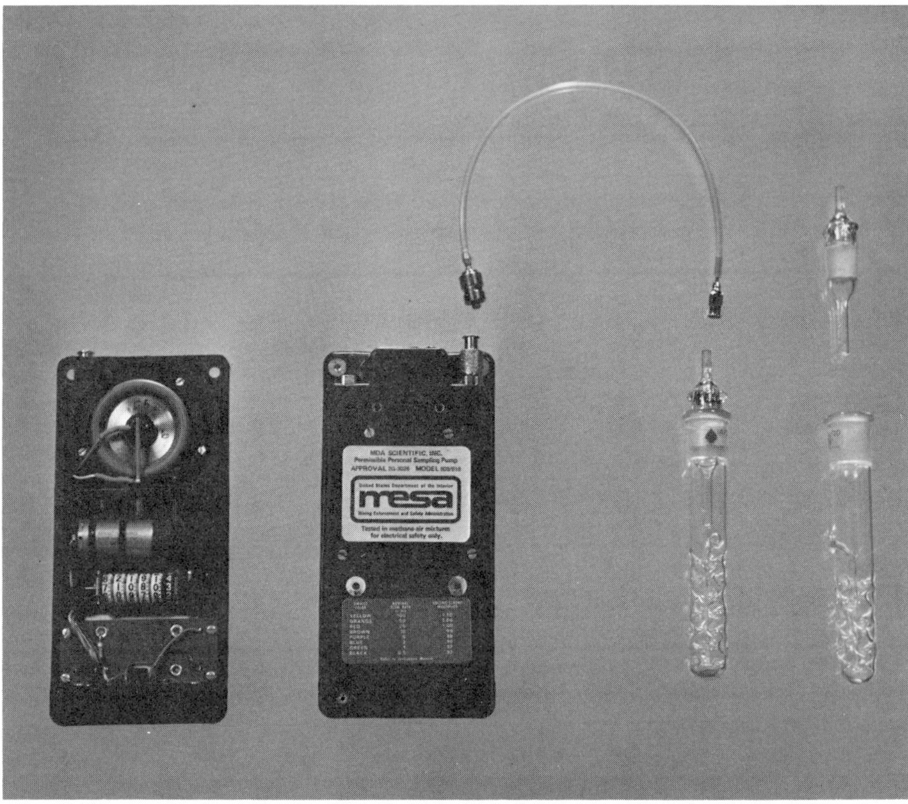

Figure 9. Personal dosimeter system

Figure 10. Dosimeter in wearing mode

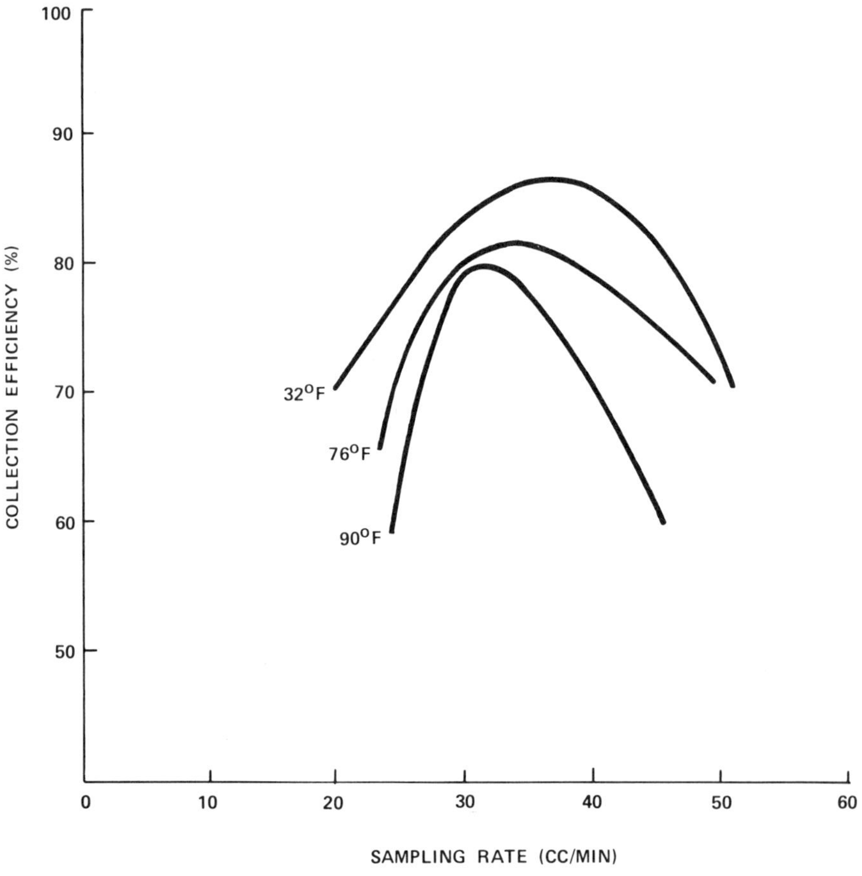

Figure 11. Bubbler influence of airflow rate on collecting efficiency as a function of temperature for low concentrations .07 to .10 ng/L

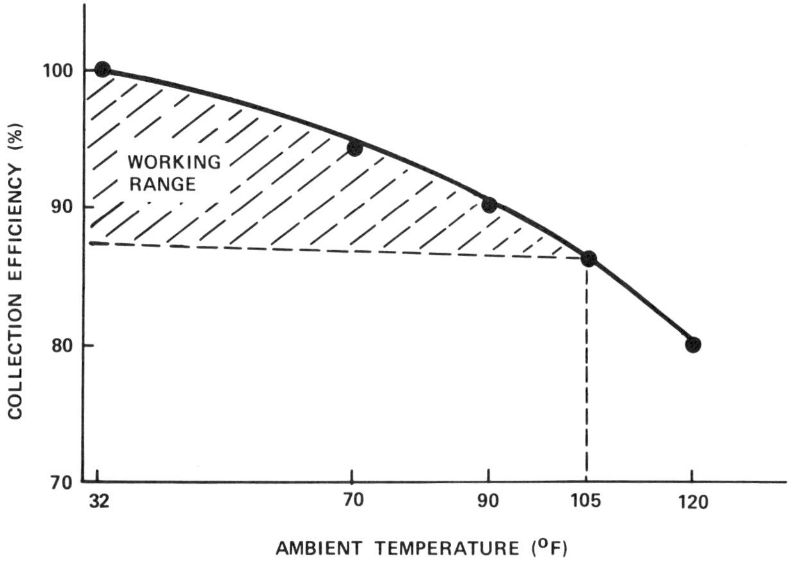

Figure 12. Bubbler average collection efficiency as a function of temperature for 30 cm³/min sampling rate and low generated concentrations

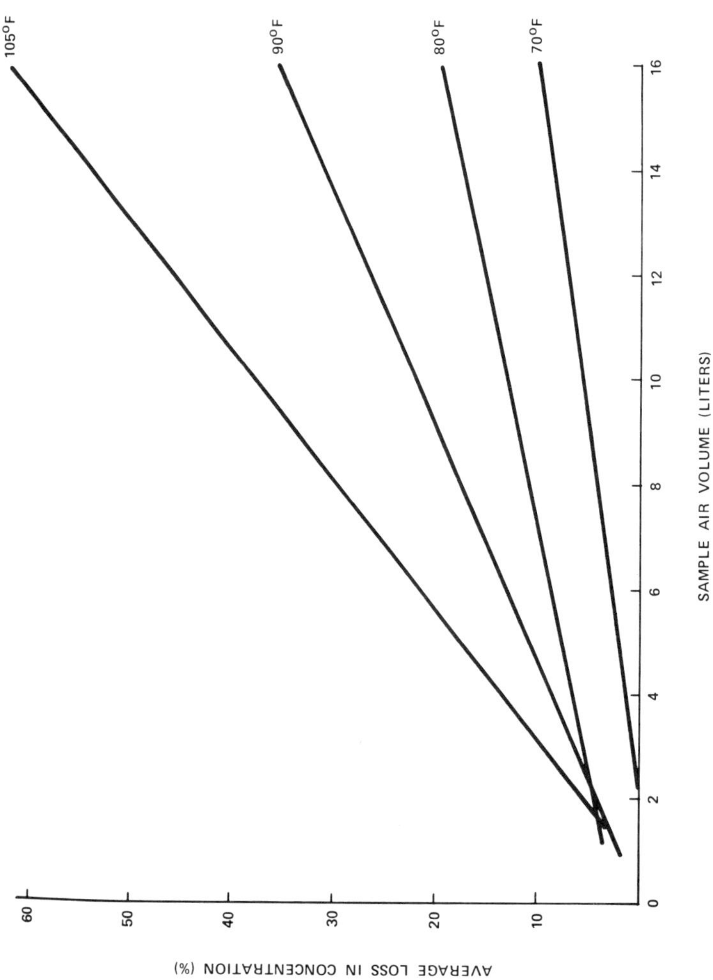

Figure 13. Desorption curves for aerated bubbler. Percent loss in concentration as a function of sample volume and temperature.

samples at a very low flow of 30 cc/min or 14.4 liters of air for an 8-hour period, which is 500 times less than the average breathing rate of a man at 15 ℓ/min. The collection efficiency for the bubbler design maximizes at approximately 30 cc/min as shown in Figure 11.

A data base for predicting the chemical behavior of the bubbler was established. The generated data base is used to characterize the sensitivity and stability of the bubbler in the presence of known quantities of toxic vapors under ambient stress conditions. The data base is used to correlate the exact relationship between field sampling results and a worker's daily exposure.

The chemical characterization work included the following essential determinations:

1. The experimental data permitted the prediction that the average collection efficiency for the aerated bubbler system will be above 90% for the temperature range sampled. This is shown in Figure 12.

2. The desorption of the bubbler was quantified as a function of sample volume and temperature as shown in Figure 13. Retention of the contaminant in the bubbler after 8 hours of aeration is sufficient for analytical evaluation.

3. The optimum pH of the absorbent was found to be in the range of 3.7 to 4.5 where the absorbed contaminant is most stable over the sampling temperatures of 32° to 105°F.

4. The colorimetric method of enzymatic analysis was found to be specific and sensitive to the levels of absorbed contaminant collected when sampling around the limit concentration of 10^{-8} g/m^3.

5. A statistical analysis indicated a high confidence level in the relevancy of the data regarding reliability and reproducibility. The system inaccuracy results from agent desorption at elevated temperatures as shown in Figure 13.

These devices are now being tested in demilitarization plants as monitors of stack exhaust atmospheres and work areas. We have not had the opportunity to determine the sensitivity of the devices to the numerous enzyme-inhibiting compounds which would be of interest to others. Generally, however, it is reasonable to expect, for instance, less sensitivity to P=S and P—S linkages and conversely better sensitivity to compounds where the sulfur is replaced by oxygen due to greater toxicity of these compounds. Such compounds as the carbamates, due to their inhibiting qualities, would also be expected to be detected with these systems.

It is expected that these systems could find application as highly sensitive monitors during manufacturing, filling, and storage of specific compounds.

RECEIVED November 30, 1978.

19

Safety Design Criteria Used for Demilitarization of Chemical Munitions

ROBERT P. WHELEN

DRC PM-DRD, Aberdeen Proving Grounds, MD 21010

This paper is intended to provide a brief historical overview of the Army's chemical demilitarization program and provide some insight to the types of disposal processes, procedures, and equipment employed.

Background

In the fall of 1968, the Department of the Army directed the disposal of certain chemical munitions which were obsolete and excess to the National deterrent stockpile. The proposed-disposal plan, designated Operation Chase, provided for these munitions to be transported across country (Denver, Colorado, to the East Coast), be loaded on excess freighter type boats and taken under US Coast Guard escort to a previously designated explosives dumping site beyond the Atlantic Continental Shelf and sunk.

This operation was suspended because of public concerns expressed over the transportation of hazardous materials through the various states, and criticism by environmentalists concerning potential impacts on marine life. In May 1969, the National Academy of Sciences (NAS) was requested by the Department of Defense to provide an assessment of the Operation Chase disposal plan. An Ad Hoc Advisory Committee of the Academy, composed of 12 experts from leading industrial, educational, and research institutions, submitted a report to Department of Defense in June 1969 from which the guidance shown on inclosure 1 has been extracted.

The obvious intent of the NAS study was to avoid sea dumping in the future by providing readily available, ecologically safe means of destroying lethal chemical munitions by current or developed techniques. Prior to 1969, the methods for disposing of obsolete or unserviceable chemical munitions were those listed on Inclosure 2. However, based on the NAS recommendations, the Army developed a program with two main objectives: First, to develop the procedures, construct the facilities/equipment, and to dispose of the chemical munitions identified as obsolete and excess for Operation Chase at the storage site, which was Rocky Mountain Arsenal, Denver, Colorado; and second, to develop a prototype facility for demilitarization of all types of chemical munitions in the stockpile as well as storage containers, including detoxification of nerve agents and mustard fills.

Subsequently, the Congress passed three Public Laws which impact on lethal demilitarization programs (Inclosure 3). In summary, the Public Laws place stringent controls on transportation of lethal chemical materiel from present storage locations, requires detoxification of chemical-biological agents prior to disposal and requires the preparation and coordination of environmental impact statements and disposal plans with various government and local agencies.

Since that time the obsolete/excess chemical munitions stored at Rocky Mountain Arsenal (RMA) and originally scheduled for disposal by sea dump have been destroyed onsite in compliance with the stringent guidance of the NAS, Department of the Army, and the Public Laws, with the exception of some bulk carbonyl chloride (phosgene) which is currently being removed by an industrial buyer. During these operations, the Army had successfully disposed of approximately 15 million pounds of lethal chemical nerve and mustard agents, in excess of 800,000 pounds of explosives, and various munitions components, including in excess of 1.6 million fuzes. Most importantly, these disposal operations have been conducted without serious injuries to personnel or insult to the environment.

In addition, the prototype demilitarization facility has been designed and constructed at Tooele Army Depot, Utah, and is currently being operated in a series of test programs. These test programs are designed to develop and refine the various demilitarization process technologies that would be required for the eventual disposal of the current stockpile of chemical munitions. The facility will also accomplish the disposal of the unserviceable munitions at the Depot.

Design Philosophy

The Army guidance on chemical demilitarization, based on the NAS recommendations, is summarized in Inclosure 4. In response to this guidance, extensive effort has been expended in the development of large scale process technology and concept designs, including explosive containment, remote mechanical process control, large volume chemical agent treatment, and advanced pollution control systems. Worker exposure standards and environmental emission standards have also been established. In addition, significant advances have been made in the "state-of-the-art" for chemical agent detection and monitoring equipment, analytical procedures, and protective clothing. (Inclosure 5) In general, demilitarization of a chemical munition involves these four steps: (a) separation of the chemical agent from explosive components; (b) detoxification of

- ASSUME ALL SUCH AGENTS AND MUNITIONS WILL REQUIRE EVENTUAL DISPOSAL AND THAT DUMPING AT SEA SHOULD BE AVOIDED.
- UNDERTAKE A SYSTEMATIC STUDY OF OPTIMAL METHODS OF DISPOSAL ON APPROPRIATE MILITARY INSTALLATIONS INVOLVING NO HAZARDS TO GENERAL POPULATION OR POLLUTION OF ENVIRONMENT.
- ADOPT TECHNIQUES SIMILAR TO THOSE USED BY THE ATOMIC ENERGY COMMISSION IN DISPOSING OF RADIOACTIVE WASTE.
- REGARD LARGE SCALE DISPOSAL FACILITIES AS A REQUIRED COUNTERPART TO EXISTING STOCKS AND PLANNED MANUFACTURING OPERATIONS.

Figure 1. Extracts from recommendations by National Academy of Sciences on disposal of chemical warfare agents and munitions

- OPEN PIT BURNING
- LAND BURIAL
- OCEAN DUMPING

Figure 2. Disposal methods prior to 1969

- 91-121 ARMED FORCES APPROPRIATION ACT OF 1970 (ESTABLISHED STRINGENT CRITERIA FOR TRANSPORTATION AND OPEN AIR TESTING OF LETHAL CHEMICAL-BIOLOGICAL AGENTS)
- 91-441 ARMED FORCES APPROPRIATION ACT OF 1971 (REQUIRED CHEMICAL-BIOLOGICAL AGENTS TO BE DETOXIFIED OR MADE HARMLESS TO MAN PRIOR TO DISPOSAL EXCEPT IN EXTREME EMERGENCY)
- 91-190 NATIONAL ENVIRONMENTAL POLICY ACT OF 1969 (REQUIRED PREPARATION AND COORDINATION OF EIS, EIA PRIOR TO CONDUCTING DEMIL OPERATIONS)

Figure 3. Public laws governing chemical demilitarization

- ABSOLUTE SAFETY AND SECURITY RATHER THAN COST OR TIME.
- MAXIMUM PROTECTION FOR OPERATING PERSONNEL.
- ABSOLUTE ASSURANCE OF TOTAL CONTAINMENT OF AGENT.
- INCONTROVERTIBLE DATA TO JUSTIFY PERSONNEL SAFETY, SECURITY, AND COMMUNITY SAFEGUARD ASPECTS.

Figure 4. Army guidance on chemical demilitarization based on NAS recomendations

the agent; (c) thermal destruction of the explosives, and (d) decontamination of the residue (Inclosure 6).

Design Criteria

The technical approach to the demilitarization programs is a total systems approach to the potential personnel safety and environmental problems associated with lethal chemical agents and explosives. The general areas of consideration are illustrated in Inclosure 7.

Total Containment and Ventilation (Inclosure 8)

The term "total" containment in this discussion means containment of the lethal chemical agent within the approved environmental emission standards. Toxic material control or minimizing contamination, is accomplished by equipment design and process techniques such as minimizing exposed surfaces during the agent draining process, employing a closed system for transfer of toxic materials and for storage tanks or reactors with process scrubbers for venting. In addition, procedures are established to chemically decontaminate agent wetted surfaces (equipment or munition components) as soon as possible in the operational sequence to minimize evaporation. Explosive containment is provided for all operations in which the explosive and/or fuze components are being processed other than by normal handling methods. An example of both explosive containment and remote control operations is illustrated in Inclosures 9-13. The facility used to dispose of the M34, 1000 pound nerve agent cluster bomb at Rocky Mountain Arsenal in Denver was the same facility used to assemble the munition in the early 1950s (Inclosure 9). Explosive containment was provided for by a two-foot thick, steel reinforced concrete building with an overhead plenum of approximately 400,000 ft^3. This facility was refurbished to satisfy the maximum credible explosive accident (MCEA) which was considered to be the detonation of an entire M34 cluster or 76 individual M125 bombs. This resulted in a design for 52 lbs of explosive and 200 lbs of nerve agent GB. The redesigned involved: blast doors, blast valves, air locks, showers-suits, and the entire ventilation system. There were over 21,000 M34 clusters as shown in Inclosure 10 in storage. The cluster contained 76 M125 bombs (Inclosure 11) with 1/2 pound of explosive and 2.6 pounds of nerve agent. The disposal process involved separating the bombs from the cluster (Inclosure 12), safing the fuze, punching the canister and draining the liquid nerve agent, cutting the explosive away from the fuze, and burning the explosive and other combustible components. The nerve agent was subsequently neutralized chemically with NaOH. Each of these operations were conducted remotely with closed circuit television observation to provide the required personnel safety. The cluster case was removed exposing the M125 bombs shown in Inclosure 13. One of the remote operations is illustrated on the next slide. That is the removal of each of the 76 bombs from the cluster by a programmed manipulator. This

- UNDERTAKE LARGE SCALE PROCESS TECHNOLOGY STUDIES AND CONCEPT DESIGNS
  - REMOTE MECHANICAL PROCESSES
  - LARGE VOLUME LIQUID TREATMENT PROCESSES
  - ADVANCED POLLUTION CONTROL SYSTEMS
- DEVELOP WORKER AND ENVIRONMENTAL STANDARDS
- DEVELOP MONITORING EQUIPMENT

Figure 5. Actions in response to guidance

- SEPARATE AGENT FROM EXPLOSIVE COMPONENTS
- CHEMICAL DETOXIFICATION OF AGENT
- THERMAL DEACTIVATION OF EXPLOSIVES
- DECONTAMINATION OF RESIDUAL MATERIAL

Figure 6. Chemical demilitarization process

TOTAL CONTAINMENT

VENTILATION

MONITORING/DETECTION

SAFETY/MEDICAL

Figure 7. Safety design criteria for demilitarization of chemical munitions

TOTAL CONTAINMENT

- TOXIC MATERIAL CONTROL
- ACCIDENTAL EXPLOSION CONTROL
- REMOTE CONTROL EQUIPMENT
- OPERATING AND MAINTENANCE PROCEDURES

VENTILATION

- VOLUME FLOW/FACE VELOCITY
- AIR LOCKS/BUFFER ZONES
- LOCALIZED VENTILATION
- FILTERS, AFTERBURNERS, SCRUBBERS

Figure 8. Safety design criteria for demilitarization of chemical munitions

Figure 9. Chemical demilitarization facility at Rocky Mountain Arsenal

Figure 10. M34, Nerve Agent GB Cluster, 1000 LB Class

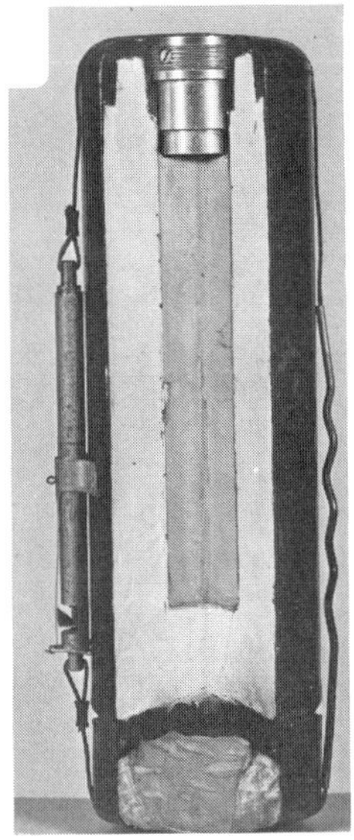

Figure 11. M125 Nerve Agent GB Bomblet

Figure 12. M34 Cluster with case removed

Figure 13. M34 Cluster dissambly programmed manipulator

Figure 14. Chemical agent munitions disposal system (CAMDS) at Tooele Army Depot

operation was controlled and monitored in a central control room utilizing CCTV. Some other types of explosive containment and remote control equipment are illustrated in Inclosures 14-19 showing portions of the prototype facility at Tooele Army Depot in Utah. The overview (Inclosure 14) illustrates the remoteness of the site approximately 60 miles southwest of Salt Lake City. Inclosure 15 shows the use of steel reinforced concrete for explosive containment in the explosive deactivation furnace operation. Inclosure 16 illustrates the use of a 10 foot diameter steel cylinder for explosive containment of the M55 Rocket demilitarization process. Inclosures 17 and 18 illustrate the M55 Chemical Rocket intact and cut into seven sections. The cutting process is accomplished with the equipment illustrated in Inclosure 19 which is housed in the explosive containment cylinder. The control room for this and other operations at the disposal site at Tooele Army Depot is shown in Inclosure 20. Provisions have been made for a remote controlled computer programmed operation including closed circuit television of all critical/hazardous operations. Another type of remote operation is illustrated in Inclosure 21. This armored personnel carrier was designed for recovery of M55 Rockets that had been burned and buried in prior operations at Dugway Proving Ground. The armored personnel carrier is equipped with a charcoal filter system and a remote control manipulator for handling the rockets that still contained chemical agent and/or explosives. The unit also includes closed circuit television cameras for observation by the operator as well as at the central control room. Some of the rockets that were recovered are illustrated in Inclosure 22. This program is also complete.

With respect to ventilation systems, we employ standard design criteria for volume flow rate, negative pressure, and face velocity requirements in contaminated and potentially contaminated areas. In contaminated plant areas, we employ a minimum of 20 air changes per hour, while in clean operating areas we use a minimum of 6 air changes per hour. All entries to the contaminated areas incorporate airlocks to prevent migration into the clean areas. In addition, localized ventilation is used in operations that have a high potential for contamination such as agent drain stations. All of our ventilation sources are processed through charcoal filters, afterburners, and/or wet chemical scrubbers. Inclosures 23-27 illustrate a typical filter installation and an afterburner-wet scrubber installation at the Tooele site.

We have 12 separate filter systems at the site ranging from 333 to 15,000 CFM (Inclosure 23). A typical application is the ADS - the facility used to destroy the nerve agents GB & VX by chemical neutralization (Inclosure 24); GB with NaOH and VX by an acid chlorination process. The actual filter system installation is shown in Inclosure 25 and has a capacity of 15,000 CFM. These filter units consist of a low efficiency particulate prefilter, a high efficiency particulate HEPA filter,

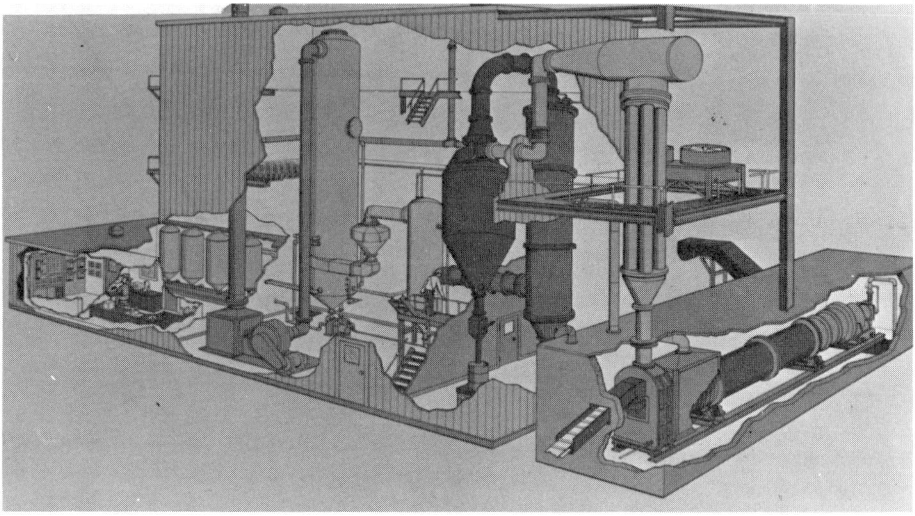

Figure 15. Explosive deactivation furnace facility

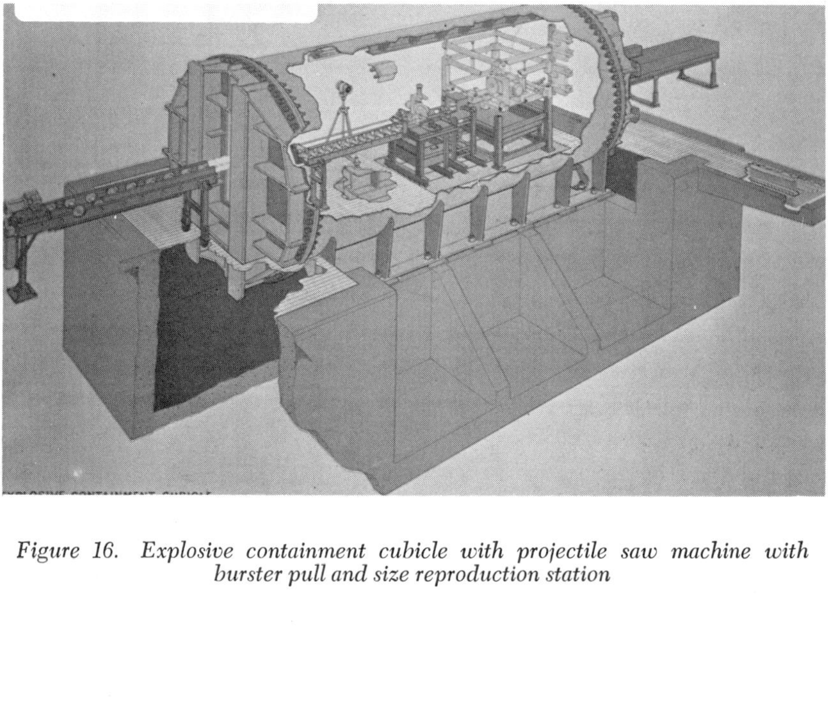

Figure 16. Explosive containment cubicle with projectile saw machine with burster pull and size reproduction station

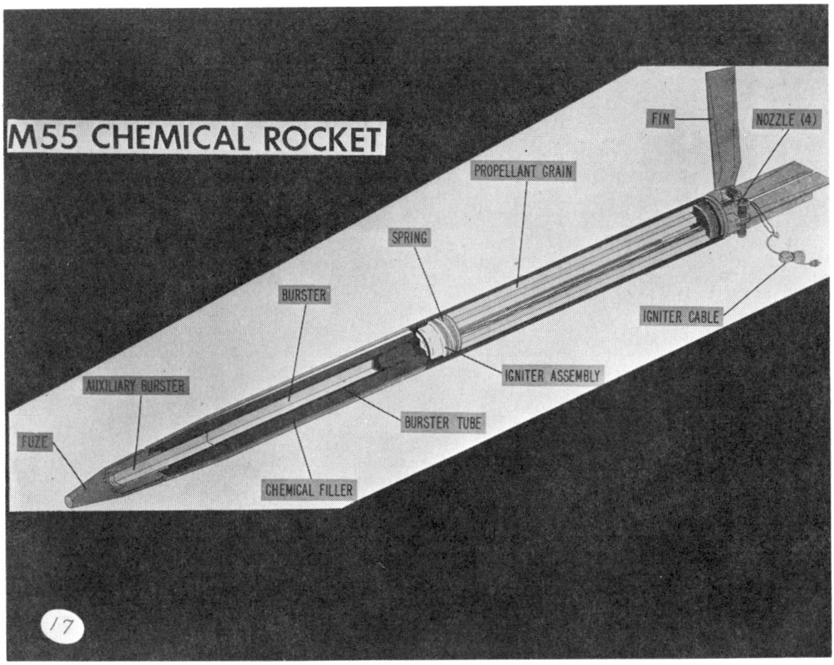

Figure 17. M55 chemical rocket

Figure 18. M55 chemical rocket—launching tube after sawing

Figure 19. Rocket demilitarization machine

Figure 20. CAMDS control system

Figure 21. Armored personnel carrier modified for chemical munition handling

Figure 22. M55 rocket recovery at Dugway Proving Ground

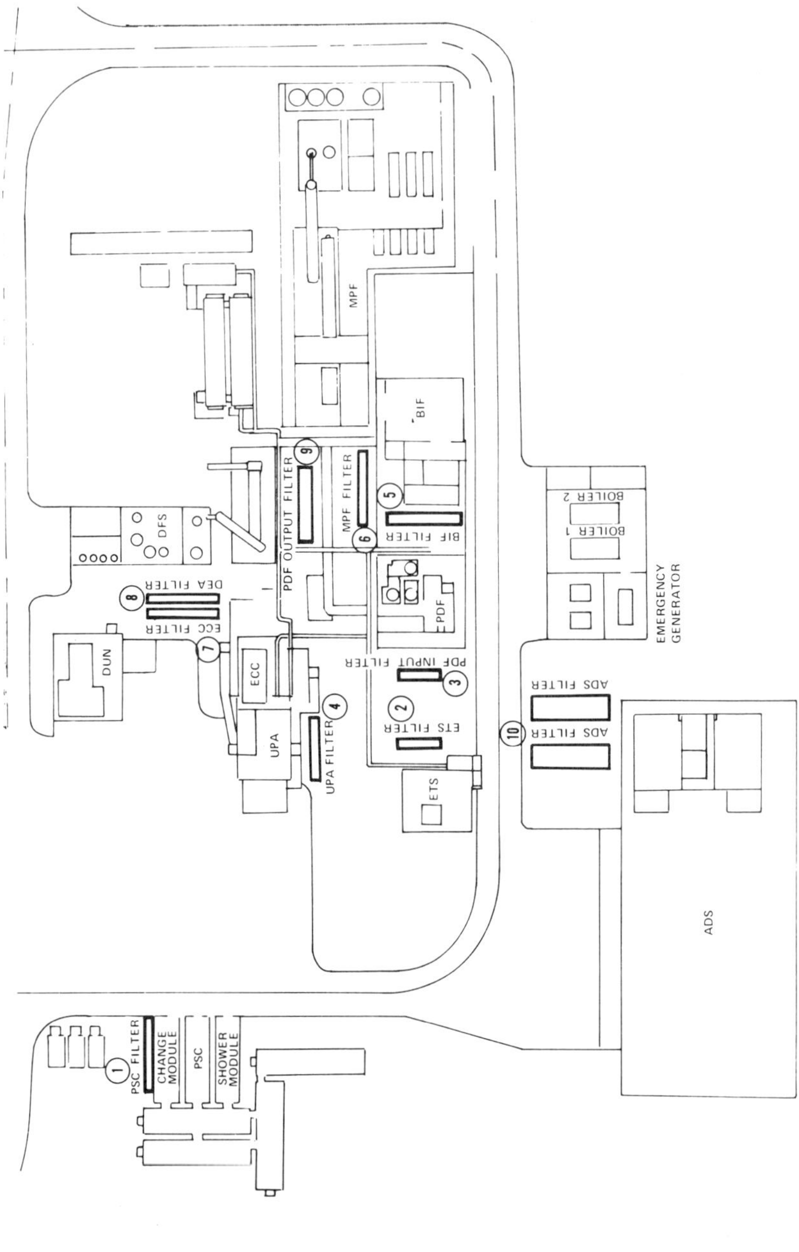

Figure 23. CAMDS site chemical filter system

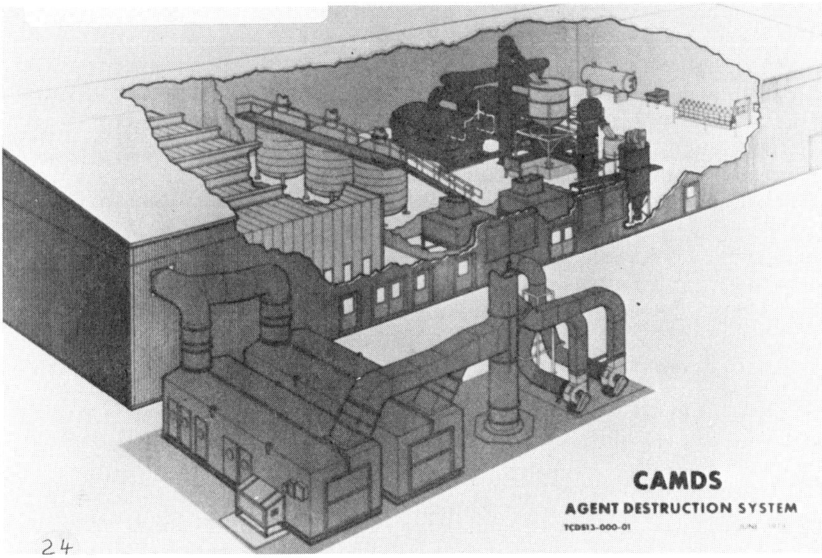

Figure 24. CAMDS chemical agent destruct facility filter system

Figure 25. Chemical filter installation at CAMDS agent destruct facility

and an activated carbon filter for agent adsorption. In order to insure total containment, we normally employ two sets of filters in series and monitor between filter banks to insure that a breakthrough is detected when the first bank is breached. Some operations dictate a separate set of double filters in parallel if we choose not to shutdown the operation while a filter change is being made. We also monitor on the stack to verify that emission standards are being met.

A typical afterburner-wetscrubber system is that used on the explosive deactivation furnace at Tooele (Inclosure 26). In this case we have a variety of combustion products from the explosive-propellent burning as well as residual chemical agent. The system employs a cyclone (primarily for fiberglass), an after-burner operating at 1600°F with a 1 second residence time, a quench, a venturi scrubber, and a packed column scrubber (pall rings) currently operating with a caustic scrubber media. The actual installation is shown in Inclosure 27.

The remaining two general areas of concern are monitoring and safety-medical aspects (Inclosure 28).

In the area of chemical monitors and detectors we employ a combination of alarms or monitors and chemical bubbler systems in the plant, on stacks, and at perimeters. Our objective is to provide adequate warning of process upsets - either major or minor that could affect workers, surrounding populations, or the environment. We have developed exposure and emission standards for each of the chemical agents being disposed of and these standards dictated that we advance the state-of-the-art for detectors and analytical procedures. The standards for GB nerve agent for example are:

3×10^{-4} mg/m^3 - stack

1×10^{-4} mg/m^3 - inplant worker TLV

3×10^{-6} mg/m^3 - amb. air qual.

During our RMA operations the only available monitoring systems were a plant alarm used in the '50's and '60's referred to as the M5 and a new field alarm referred to as the M8 which can detect 0.1 - 0.3 mg/m^3 within one minute. The response time is more than adequate, however, the detection level is 3 logs above the 1×10^{-4} inplant TLV. So we used these alarms for upset conditions and employed chemical bubblers with a two hour sampling time to detect at the 0.0001 level. The bubbler however is cumbersome, requires refrigeration during sampling and with the time for analysis, has a turnaround time of 3-4 hours depending on the number of bubblers being processed. The RMA workload was substantial in that a 3 shift, 7 day work schedule was maintained for 3 years with in excess of 13,000 bubblers/month processed. Since that time we have developed an automated enzymatic detector that is capable of detecting the

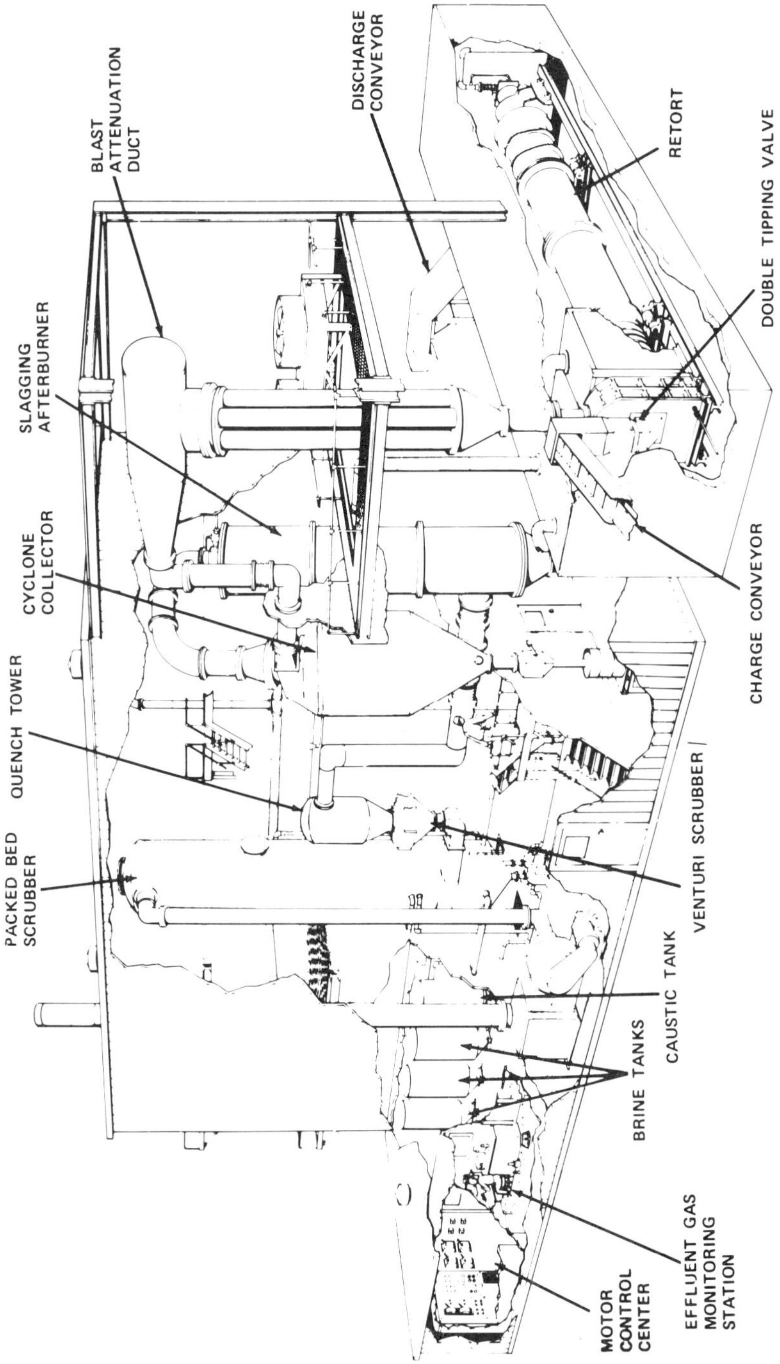

Figure 26. Afterburner—wet scrubber system at CAMDS explosive deactivation facility

Figure 27. CAMDS explosive deactivation facility

MONITORING AND DETECTION

- IN-PLANT
- EXHAUST STACKS
- PERIMETER

SAFETY/MEDICAL

- PROCEDURES/TRAINING
- PROTECTIVE CLOTHING
- PHYSICAL EXAMINATIONS/PERIODIC CHECKUPS
- AID STATION, MEDICAL TECHNICIANS, AMBULANCES

Figure 28. Safety design criteria for demilitarization of chemical munitions

inplant TLV of 1 x 10^{-4} mg/m^3 in 8-10 minutes.

Inclosure 29 depicts the general detector plan for the Tooele plant with a combination of bubblers and chemical alarms located in all work areas, air locks, and stacks.

Inclosure 30 shows the M55 GB Nerve Agent Rocket unpack area. In this area the individual rockets are removed from their packing of 15 rockets. The rockets are transported in the steel container shown in the center of the room. This operation is monitored by one of the new alarms - referred to as an RTM for Real Time Monitor and two bubbler stations. The detection systems are positioned to provide as representative a sample as possible (with a point sampling system). In addition one of the bubblers samples the room exhaust. The RTM is being evaluated in the Tooele operation and if successful could replace some of the bubbler requirements.

We use the same type of detectors on our exhaust stacks, however, the specific applications and sensitivity levels have to be tailored due to interferences by various combustion products.

Perimeter monitoring stations are also employed during our operations. At the Tooele site we have 8 stations around the perimeter of the disposal plant (Inclosure 31). This is the South Area of Tooele Army Depot. Inclosure 32 illustrates one of the individual stations which monitor for CO_2, NO_2, total oxidants, particulate; chemical agent plus wind speed and direction.

Our safety and medical programs are designed to supplement the plant and equipment designs. Extensive operations, maintenance, and emergency training is accomplished with inert munitions and explosively configured munitions with simulant chemical fill before full scale chemical agent operations are started. Detailed procedures are developed, reviewed, and practiced for each operation by all personnel including classroom as well as "hands-on" training. Developing procedures involves identifying the maximum hazards, making every effort to minimize these potential hazards and developing the procedures accordingly. One of our primary areas of concern is the protective clothing used - particularly the level A clothing used for maximum protection in areas of known or having a high potential for being contaminated.

During the RMA operations we used the Army's field type, (M3) level A clothing shown in Inclosure 33. It is basically a butyl rubber suit designed for protection against liquid contamination and uses the M9 protective mask with carbon filter for vapor protection. This suit has been used extensively over the years, however, it does not satisfy the OSHA requirement for an air supplied suit. Based on this requirement we developed a new air supplied suit (Inclosure 34) and are in the final stages of testing prior to use at Tooele. This suit incorporates a reusable respirator with an emergency self-contained air supply

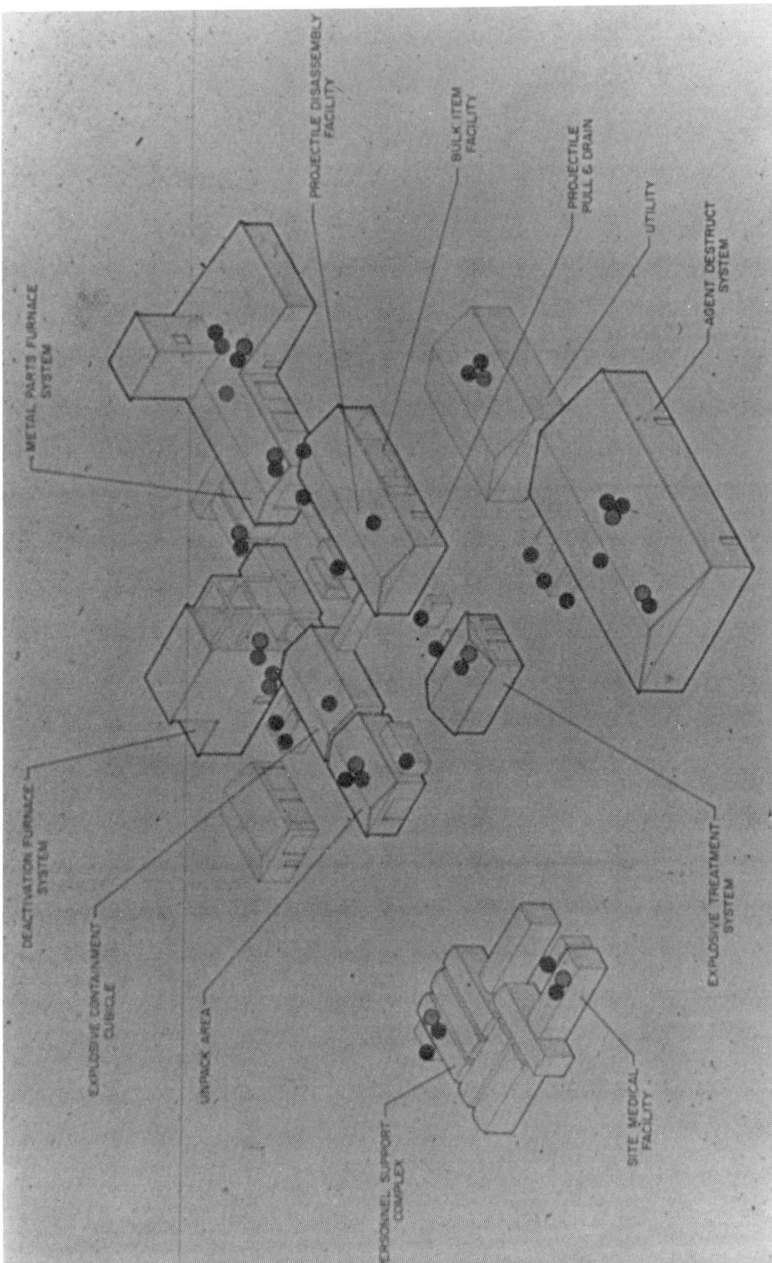

Figure 29. CAMDS chemical agent monitoring—detector plan

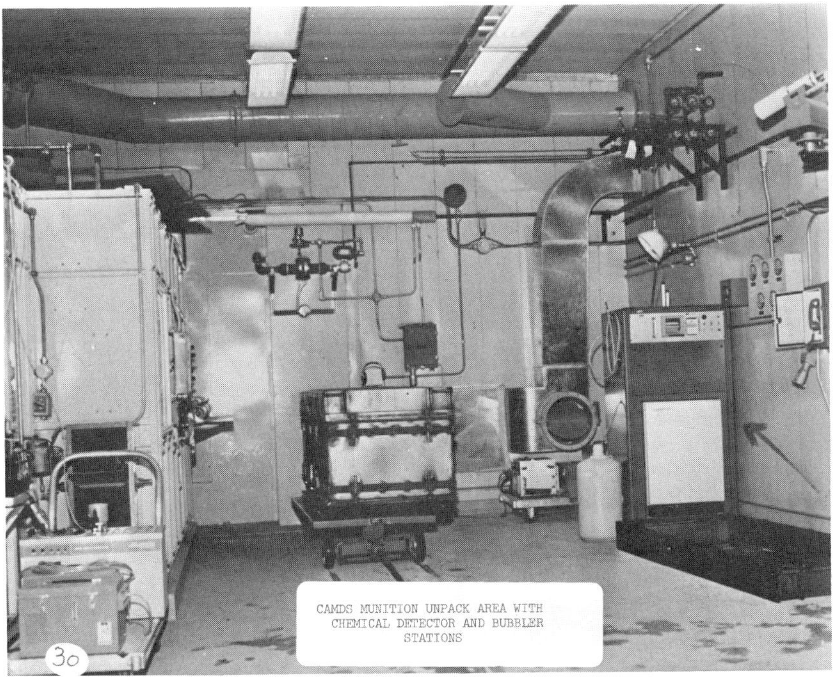

Figure 30. CAMDS munition unpack area with chemical detector and bubbler stations

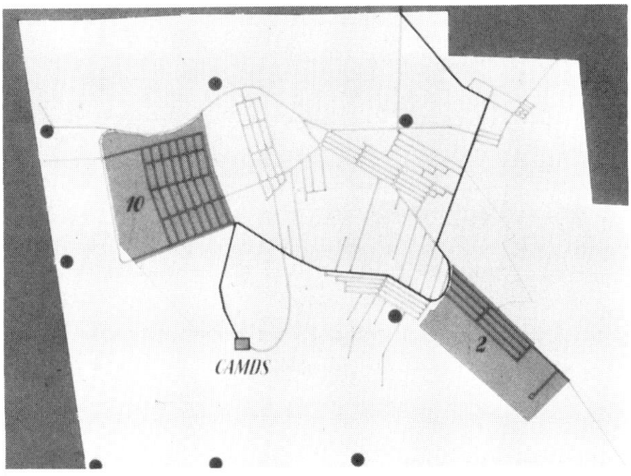

Figure 31. Perimeter monitoring stations. South area.

Figure 32. Perimeter monitoring station at CAMDS

Figure 33. M3 protective suit

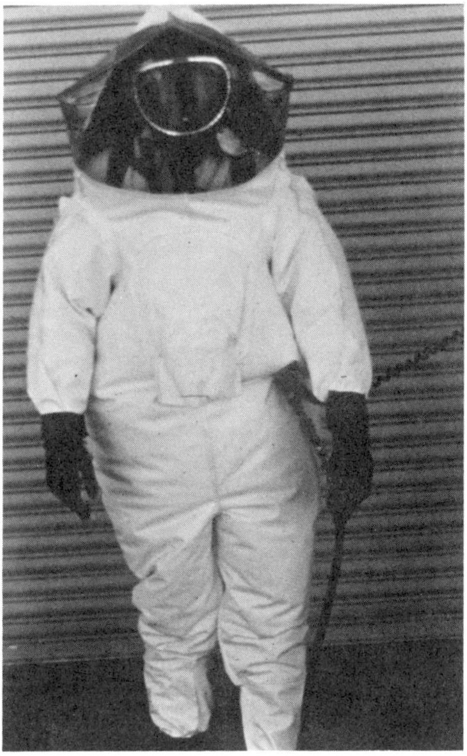

Figure 34. New chemical agent protective suit

Figure 35. New chemical agent protective suit

and a disposable outergarment made of two-ply chlorinated polyethylene (Inclosure 35).

With respect to the medical aspects, we develop a detailed medical plan for each program that is coordinated through the Army's medical community and DHEW. The plans include:

(1) Preassignment physicals for all personnel.

(2) Physician training in chemical agent treatment.

(3) Personnel training to recognize symptoms every 6 months.

(4) Special attention to red blood cell cholinesterese. Surveillance - baseline, periodic checks.

(5) Emergency aid station, ambulances, etc.

(6) Physicians and/or trained medical technicians onsite during all operations.

As stated at the outset, this paper is intended to provide a historical overview and some insight as to the types of operations, procedures and equipment used in the Army's Demilitarization Program for Chemical Munitions. In each of the areas discussed there has been a considerable amount of study, laboratory experimentation and pilot testing accomplished to define the specific design criteria.

RECEIVED November 22, 1978.

INDEX

INDEX

A

B

C

T

U

V

W

Z